WEB
SOCIAL
SCIENCE

For Kazuko

WEB
SOCIAL
SCIENCE

CONCEPTS, DATA AND TOOLS FOR
SOCIAL SCIENTISTS IN THE DIGITAL AGE

ROBERT ACKLAND

Los Angeles | London | New Delhi
Singapore | Washington DC

Los Angeles | London | New Delhi
Singapore | Washington DC

SAGE Publications Ltd
1 Oliver's Yard
55 City Road
London EC1Y 1SP

SAGE Publications Inc.
2455 Teller Road
Thousand Oaks, California 91320

SAGE Publications India Pvt Ltd
B 1/I 1 Mohan Cooperative Industrial Area
Mathura Road
New Delhi 110 044

SAGE Publications Asia-Pacific Pte Ltd
3 Church Street
#10-04 Samsung Hub
Singapore 049483

Editor: Chris Rojek
Editorial assistant: Martine Jonsrud
Production editor: Katherine Haw
Copyeditor: Richard Leigh
Proofreader: Jonathan Hopkins
Indexer: Cathryn Pritchard
Marketing manager: Alison Borg
Cover design: Lisa Harper
Typeset by: C&M Digitals (P) Ltd, Chennai, India
Printed and bound by CPI Group (UK) Ltd

Library of Congress Control Number: 2012950465

British Library Cataloguing in Publication data

A catalogue record for this book is available from
the British Library

ISBN 978-1-84920-481-1
ISBN 978-1-84920-482-8 (pbk)

Contents

List of Figures   ix
List of Tables   x
List of Boxes   xi
About the Author   xii
Preface   xiii

1   Introduction   **1**
   1.1   The web: technology, history and governance   1
   1.2   Examples of online computer-mediated interaction   5
   1.3   Cyberspace, virtual communities and online social networks   7
         1.3.1   Cyberspace   8
         1.3.2   Virtual communities   10
         1.3.3   Online social networks   12
   1.4   Disciplinary approaches to researching the web   13
   1.5   Construct validity of web data   16
   1.6   Shaping force or social tool?   16
   1.7   Conclusion   17

I   WEB SOCIAL SCIENCE METHODS   **19**

2   Online research methods   **21**
   2.1   Dimensions and modes of online research   21
   2.2   Online surveys   25
         2.2.1   Sampling: basics   26
         2.2.2   Types of Internet surveys   27
         2.2.3   Online surveys: process and ethics   28
         2.2.4   Online survey example: election studies and
                 election polls   29
         2.2.5   Other issues   30
   2.3   Online interviews and focus groups   31
         2.3.1   Types of online interviews   31
         2.3.2   Online interviews: process and ethics   32
         2.3.3   Online focus groups   33
         2.3.4   Other issues   34
   2.4   Web content analysis   35
         2.4.1   Quantitative web content analysis   35
         2.4.2   Qualitative web content analysis   38
         2.4.3   Web content used in data preparation   40

2.5   Social media network analysis   40
2.6   Online experiments   40
      2.6.1   Online laboratory experiments   40
      2.6.2   Online field experiments   41
      2.6.3   Online natural experiments   41
2.7   Online field research   41
2.8   Digital trace data: ethics   43
2.9   Conclusion   46

3   Social media networks   **48**
3.1   Social networks: concepts and definitions   48
      3.1.1   An example school friendship network   51
3.2   Social network analysis   55
      3.2.1   Social relations and social networks   55
      3.2.2   Statistical analysis of social networks   58
3.3   Social media networks   61
      3.3.1   Representing online interactions as interpersonal
              networks   61
      3.3.2   Threaded conversations   65
      3.3.3   Social network sites   69
      3.3.4   Microblogs   72
3.4   Social networks, information networks and
      communication networks   73
      3.4.1   Flows of information and attention   74
3.5   SNA metrics for the example school friendship
      network (advanced)   75
      3.5.1   Node-level SNA metrics   75
      3.5.2   Network-level SNA metrics   76
3.6   Conclusion   77

4   Hyperlink networks   **78**
4.1   Hyperlink networks: background   78
      4.1.1   Motives for sending, and benefits
              of receiving, hyperlinks   79
      4.1.2   Hyperlink network nodes, ties and boundaries   80
4.2   Three disciplinary perspectives on hyperlink networks 82
      4.2.1   Citation hyperlink networks   82
      4.2.2   Issue hyperlink networks   83
      4.2.3   Social hyperlink networks   83
      4.2.4   Comparing the disciplinary perspectives   84
4.3   Tools for hyperlink network research   86
      4.3.1   Web crawlers   86
      4.3.2   Historical web data   91
      4.3.3   Blogs   92
4.4   Conclusion   94

II WEB SOCIAL SCIENCE EXAMPLES   **95**

5 Friendship formation and social influence   **97**
  5.1   Homophily in friendship formation   97
       5.1.1   Measurement issues   97
       5.1.2   Friendship formation in Facebook   99
       5.1.3   Online dating   101
  5.2   Social influence   103
       5.2.1   Identifying social influence   103
       5.2.2   Social influence in social media   105
  5.3   Conclusion   110

6 Organisational collective behaviour   **111**
  6.1   Collective behaviour on the web: background   111
  6.2   Collective action and public goods   113
       6.2.1   Hyperlink networks as information public goods   113
  6.3   Networked social movements   114
  6.4   Conclusion   118

7 Politics and participation   **119**
  7.1   Visibility of political information   119
       7.1.1   Power laws and politics online   119
  7.2   Social and political engagement   123
       7.2.1   Web use and social capital   123
       7.2.2   Political engagement   127
  7.3   Political homophily   129
       7.3.1   Divided they blog   130
  7.4   An introduction to power laws (advanced)   132
  7.5   Conclusion   136

8 Government and public policy   **138**
  8.1   Delivery of information to citizens   138
       8.1.1   Government hyperlink networks   139
  8.2   Government authority   142
       8.2.1   Civil unrest   143
       8.2.2   Internet censorship   144
  8.3   Public policy modelling   146
       8.3.1   The mapping principle   146
       8.3.2   The macroeconomics of a virtual world   147
  8.4   Conclusion   147

9 Production and collaboration   **149**
  9.1   Peer production and information public goods   149
       9.1.1   Peer production   150
       9.1.2   Information public goods   152

9.2 Scholarly activity and communication 156
   9.2.1 Webometric measures of scholarly output
      and impact 156
   9.2.2 Reconfiguring access to scholarly information
      and expertise 158
9.3 Network structure and achievement 159
   9.3.1 Identifying a 'network effect' in outcomes 159
   9.3.2 Structural holes in Second Life 160
9.4 Conclusion 162

10 Commerce and marketing **163**
10.1 Distribution of product sales 163
   10.1.1 Power laws and superstars 164
   10.1.2 Evidence for the Long Tail 166
10.2 Influence in markets 168
   10.2.1 Referrals from friends 169
   10.2.2 Ratings systems 171
   10.2.3 Recommender systems 173
10.3 Conclusion 173

References 175
Index 191

List of Figures

3.1	School friendship network	51
3.2	1.0-degree ego network	53
3.3	1.5-degree ego network	54
3.4	2.0-degree ego network	54
3.5	Transitive triad extracted from school friendship network	59
3.6	Threaded conversation network	62
3.7	Web 1.0 network	63
3.8	Wiki network	64
3.9	Facebook network	64
3.10	Twitter network	65
3.11	Posts in a discussion topic	67
3.12	Posts in a question-and-answer topic	67
3.13	Discussion topic threads represented as reply network and top-level reply network	68
7.1	Hyperlink network of pro-choice (white) and pro-life (grey) websites (Ackland and Evans, 2005)	131
7.2	Indegree–rank plot for demonstration data	133
7.3	CDF for demonstration data	134
7.4	1 − CDF for demonstration data	135
7.5	1 − CDF for demonstration data, log–log plot	135
7.6	1 − CDF for 2005 Australian web, log–log plot	136
9.1	Exemplary authorline for answer person, Welser et al. (2007)	154
9.2	Exemplary authorline for discussion person, Welser et al. (2007)	155
9.3	Exemplary ego network for answer person, Welser et al. (2007)	155
9.4	Exemplary local network for discussion person, Welser et al. (2007)	156
10.1	Indegree–rank plot of demonstration data, with Long Tail shown	164
10.2	Probability density function for demonstration data	165
10.3	Probability density function − normal distribution	166

List of Tables

2.1 Modes of online research   25
3.1 Adjacency matrix (partial) for student friendship data   52
3.2 Edge list (partial) for student friendship data   52
7.1 Summary of power law example data   134
9.1 Comparing governance structures: market, hierarchy and
     bazaar (Demil and Lecoq, 2006)   151

List of Boxes

1.1 Resources on the web 3
1.2 Web timeline 4
1.3 Phases in the evolution of the web 4
2.1 Mechanical Turk 42
3.1 Key network definitions 49
3.2 Types of social networks 50
3.3 The strength of weak ties 57
3.4 Structural holes in social networks 58
3.5 Extracting data from Usenet: Netscan and SIOC 69
4.1 HTML and RDF 88
4.2 International Internet Preservation Consortium 92
5.1 Homophily in a US school friendship network 100
6.1 International hyperlink networks 115
6.2 Offline characteristics of NGOs and hyperlink networks 117
7.1 Political parties and the normalisation thesis 121
7.2 How power laws develop (preferential attachment) 122
7.3 Definitions of social capital 124
7.4 Who is the 'average' online gamer? 125
7.5 Internet use and political engagement: estimation approaches 128
7.6 Power laws in the real world 132
8.1 Peer-produced digital currency – Bitcoin 139
9.1 What is open source software? 150
10.1 Economics of superstars 167

About the Author

Robert Ackland is an Associate Professor in the Australian Demographic and Social Research Institute at the Australian National University. He has degrees in economics from the University of Melbourne, Yale University (where he was a Fulbright Scholar) and the ANU, where he gained his PhD in 2001. Prior to commencing his PhD, which was on index number theory in the context of cross-country comparisons of income and poverty, Robert worked as a researcher in the Australian Department of Immigration and an economist in the Policy Research Department at the World Bank, based in Washington, DC, 1995-97. Since 2002 Robert has been working in the fields of network science, computational social science and web science, with a particular focus on quantitative analysis of online social and organisational networks. He has given over 50 academic presentations in this area and his research has appeared in journals such as the *Review of Economics and Statistics*, *Social Networks*, *Computational Economics*, *Social Science Computer Review* and the *Journal of Social Structure*. Robert leads the Virtual Observatory for the Study of Online Networks project (http://voson.anu.edu.au), and teaches on the social science of the Internet, statistics, and online research methods. He has been chief investigator on four Australian Research Council grants, and in 2007 he was both a UK National Centre for e-Social Science Visiting Fellow and James Martin Visiting Fellow at the Oxford Internet Institute. In 2011, he was appointed a member of the Science Council of the Web Foundation's Web Index project.

Preface

This book aims to provide students, researchers and practitioners with the theory and methods for understanding the web as a socially constructed phenomenon that both reflects social, economic and political processes and, in turn, impacts on these processes.[1] Specifically, readers of this book will:

- learn about relevant data, tools and research methods for conducting research using web data;
- gain an understanding of the fundamental changes to society, politics and the economy that have resulted from new information and communication technologies such as the web;
- learn how Internet data are providing new insights into long-standing social science research questions;
- understand how social science can facilitate an understanding of life in the Internet age, and how approaches from other disciplines can augment the social scientist's toolkit.

There are three main motivations for social scientific research into behaviour on the web. First, it can be argued that behaviour on the web is a unique cultural form that deserves to be documented and understood. This is more a perspective taken in virtual ethnography, for example, and it is not central to the present book. Second, it may be the case that some behaviour on the web is similar enough to offline behaviour, that its study can provide new insights into the offline behaviour. This perspective is a strong one in this book. Finally, social science research into web behaviour has been motivated by the need to understand whether certain online behaviour may have effects in the 'real world', that is, that there may be a *social impact* of the web. This motivation is again an important aspect of this book.

Since the early days of the Internet there has been research into its social, political and economic impacts, with contributions from a range of disciplines: media and cultural studies, communications, economics, political science, sociology, law and public policy. So what does this book offer that is new or different? What sets web social science apart from other approaches for studying the web?

---

[1] The inventor of the World Wide Web, Tim Berners-Lee, advocates the use of 'Web' when referring to the proper noun (see http://www.w3.org/People/Berners-Lee/FAQ.html#Spelling), but for convenience, this book uses 'web' throughout.

SHAPING FORCE OR SOCIAL TOOL?

On the face of it, it would appear that the Internet has had a huge influence on the way we live our lives – it would seem that it has transformed the way we work, collaborate, engage in commerce, participate in the political process and interact socially. However, it is useful to keep in mind the words of the social historian Claude S. Fischer:

> Visions of new technologies revolutionizing the way we live are often bold, sweeping, and millenarian. They are exciting to hear; they sell books; they can earn one a good living on the corporate lecture circuit. But their shelf-life is roughly equivalent to that of a 'Big Mac' ... we ought to think more about these technologies as tools people use to pursue their social ends than as forces that control people's actions. (Fischer, 1997: 115)

This is not to say that the Internet has had *no* social impacts, but rather that the real-world impacts of the Internet are likely to be complex and hard to disentangle from other major forces, for example demographic and economic, that are affecting patterns of communication and community.

So if we agree with Fischer's view (in 1997, note) that the Internet has not controlled people's actions but has rather provided tools to allow people to pursue their social ends, the implication is that the Internet can provide new data sources for studying human behaviour. As a tool for communication, the web differs from the telephone in that interactions between people often lead to digital trace data that can be used for research: email repositories, archives of posts on forum sites and blog sites, and profiles in social network sites such as Facebook.

RECOGNISING THE DISCIPLINARY APPROACHES TO STUDYING THE INTERNET

One of the major aims of this book is to identify the disciplinary approaches to studying the web: what does a social science approach to studying the web involve, and how is it different from (or similar to) other disciplinary approaches? What is included under the banner of *social science*? The book identifies social science by the core fields of economics, politics and sociology, with a focus on research that is quantitative and/or grounded in network theory and methods.

When applied physicists look at the web, they see a massive network of web pages and the hyperlinks between them – a source of data for measuring and understanding large-scale network properties such as power laws. For information scientists, the web represents a huge scientific citation network, ripe for the application of bibliometric techniques to understand scholarly output and impact. Media and communications scholars are interested in the web as a channel for distributing news and opinion; they study the production and consumption of web content in the context of understanding opinion leadership and agenda setting. When social scientists look at the web, they see individuals and organisations interacting socially, economically and politically; often this behaviour can be described as occurring in networks.

All of these disciplines have made important contributions, and the question that needs to be asked is: is there a need for web social science? Do we need to – by using the term *web social science* rather than, say, *Internet research* or *Internet studies* – effectively establish a boundary or demarcation line that will include some disciplines and exclude others? The use of the term *web social science* does not necessarily involve boundary marking: there are a lot of people who study the Internet who do not identify themselves as social scientists, and, similarly, there is a lot of research into the Internet which, while focusing on social aspects of the Internet, is not recognisable to social scientists as being social science. The mere fact that the behaviour being studied is social does not make it sociology.

Why in this era of interdisciplinarity (encouraged by universities and funding councils) should there be a book seeking to promote the social science of the Internet? Is this not simply helping to maintain the disciplinary silos, rather than break them down? Even though the Internet was built by engineers and there has been a lot of influential work by applied physicists and computer scientists, it is important for social scientists to take an active role in the development of new approaches to studying the Internet. Otherwise they may be bound by other disciplines' tools, frameworks and research questions, and thus constrained by the world view of other academic fields.

So, while this book draws from these other disciplines, the primary goal is the advancement of social science. While future advances in our understanding about the digital and real worlds will continue to involve the collaboration of social scientists and other disciplines (as the emerging field of computational social science promises; see, for example, Lazer et al., 2009), social scientists still need to be able to make contributions on their own terms, and this book aims to equip them to do so.

THE ROLE OF NETWORKS

In the above characterisation of how the web is viewed by scholars from physics, information science, media studies and social science, the word *network* appeared several times. This book emphasises quantitative network research methods (in particular, social network analysis). Already hugely influential in disciplines such as applied physics, network science is said to be the science of the twenty-first century. While sociologists have been studying social networks for 40 years, the use of networks in other social science disciplines is still limited, often being used in a metaphoric or heuristic sense but without formal rigour or empirics.

But to understand a lot of interesting behaviour on the web you have to understand networks. This book reflects the increasing use of formal network concepts and methods to understand the structure of the web from a social science perspective. However, it is not assumed that readers are already familiar with network theory or methods: the book provides an introduction to network theory and methods for social science web research. Further, it should be noted that not all chapters in the book are focused on research involving networks. While just about any interaction between individuals and organisations can be represented as a network, often that behaviour can be more appropriately described and understood without using network concepts.

THE ROLE OF QUANTITATIVE RESEARCH

This book also emphasises the role of quantitative research in the social science of the Internet. A particular focus is on digital trace data (data that are unobtrusively collected from the Internet); however, the book also covers quantitative analysis of obtrusive online data (e.g. online surveys) as well as (to a lesser extent) offline surveys of households and individuals.

CAUSALITY VERSUS CORRELATION

Identifying causal relationships is one of the main challenges for empirical social science. The web offers opportunities to overcome methodological limitations in social science by enabling, for example, natural and field experimentation that can help discern causation and would be impossible to conduct in the real world. But the movement of social life onto the web, and the concomitant realisation that social networks are key to much of the behaviour that is of interest to social scientists, have exposed us to new methodological challenges with regard to identifying causality. The issue of causality arises in several chapters of this book and is one of the core methodological challenges for web social science.

CONSTRUCT VALIDITY OF WEB DATA

Social scientists are interested in the web as a source of data that can potentially provide new insights into long-standing questions in social science. But online relationships and social behaviours may not have quite the same meaning, or dynamics, as they do in the real world. The fact that Facebook decided to use the term 'friending' to describe the act of two people making a connection on the site does not necessarily mean that Facebook data are appropriate for studying homophily (the idea that people become friends with others who share similar attributes, or 'birds of a feather flock together'), for example. This book emphasises the importance of establishing that there is an appropriate *mapping* from the online to the offline, or, in other words, that web data have *construct validity* for the context in which they are being used.

THE WEB AS RECONFIGURING FORCE

The web has been described as a force that can 'reconfigure' or radically alter the status quo in various areas in the real world. This book considers the arguments and evidence relating to how the web can potentially: enable non-government organisations to leapfrog government and commercial interests in engaging with the public; level the political playing field in favour of minor or fringe political actors; enable protesters to more effectively circumvent government authority; democratise access to scientific expertise; or increase sales diversity as firms profit from the Long Tail.

SAMPLING IS STILL IMPORTANT

Sampling from a target population is one of the core techniques of empirical social science. In a research environment such as the web, where there are potentially millions of observations in the target population, one would think that devising appropriate sampling techniques for digital trace data would be at the forefront of the minds of methodologists. A lot of social science research into the web requires greater understanding about the units of observation (be they websites, Twitter users, Facebook users) that can be garnered via automatic methods alone. But oddly, sampling has not been a large focus of methods development for web research, and one wonders to what extent this has been due to the influence of other disciplines where a research design is not considered exciting or worthwhile unless it is viable at 'data scale'. The fact that numerous web data are in the form of networks (hence exhibiting interdependence) makes sampling less straightforward than in some real-world social science settings where observations can be validly assumed to be independently distributed of one another. However, this book emphasises the fact that sampling is an integral part of web social science.

BRINGING IT ALL TOGETHER

Finally, this book is different in that it aims to bring all these elements into a single text that equips readers with the tools, theory and methods for conducting web social science.

While the primary audience is social scientists, the book will appeal to a range of disciplines. The quantitative/mathematical nature of parts of the book will draw scientists and engineers interested in learning how social scientists are studying the web. The web is now an important tool for organisational marketing and communication and coordination (e.g. non-government organisations building virtual communities for collective action, and corporations wishing to measure their online brand presence). So this book will also be useful for practitioners in many areas.

The author has given over 50 academic presentations in this area since 2002 and a number of times has heard audience members say that they had never thought of the Internet as a research area or source of data, and that the presentation had opened up new possibilities. The major goal of this book is to open eyes both to the Internet as a source of new data for social science research, and to the insights and tools that social science provides for understanding life in the Internet age.

STRUCTURE OF THE BOOK

The book consists of an introductory chapter (Chapter 1), which introduces web social science and the major themes covered in the rest of the book, followed by nine chapters arranged in two parts.

Part I is about how to do web social science. It introduces readers to the methods, tools and data for researching behaviour on the web and for using the Internet as a delivery medium for online research tools (e.g. for studying offline behaviour via the web). It consists of three chapters:

- Chapter 2 is an introduction to online research methods.
- Chapter 3 introduces social media network analysis.
- Chapter 4 is focused on the analysis of hyperlink networks.

Part II provides examples of web social science that draw on the methods, data and tools introduced in Part I. The examples illustrate how social science has been advanced using data from the web and the contribution of empirical social science to greater understanding of the social, economic and political impacts of the Internet.

- Chapter 5 looks at homophily and the closely related topic of social influence, showing the inherent difficulty in identifying these phenomena and how web data are providing new insights.
- Chapter 6 is focused on organisational collective behaviour on the web – in particular, the hyperlinking behaviour of non-government organisations.
- Chapter 7 looks at the potential influence of the web on politics from the perspective of the visibility of political information and the engagement of citizens in the political process, and the tendency of people to cluster

on the basis of political affiliation. The chapter also covers the impact of the Internet on social connectivity more generally.

- Chapter 8 is concerned with government in the digital age, looking at how web data can be used to conduct comparative analysis of the information provision activities of government, the impact of social media on government authority and the potential of virtual worlds for public policy research.
- Chapter 9 looks at how the Internet has transformed production and collaboration, and our ability to study and measure these activities. Internet-enabled peer production and its relation to information public goods is discussed. The chapter also investigates how the web has changed our ability to measure scholarly output, and the distribution of scholarly visibility and authority. Finally, the use of data from virtual worlds for studying social networks and individual performance is covered.
- Chapter 10 is concerned with Internet marketing and commerce, first reviewing the concept of the Long Tail and related empirical evidence, and then looking at both how Internet marketing is leveraging social networks and the potential impact of online recommender systems on sales diversity.

ACKNOWLEDGEMENTS

I would like to thank Heather Crawford, Francisca Borquez and Susannah Sabine for assistance with editing and proofreading various versions of the manuscript. I would also like to thank Paul Vogt for helpful comments on an earlier version of the manuscript, and Chris Rojek, Martine Jonsrud and Katherine Haw from SAGE for helping me to write this book.

I also thank my students in the Social Science of the Internet stream of the Master of Social Research and PhD programme at the Australian National University, who have contributed much to my thinking about this topic, and who have been reading very underdone versions of chapters since 2008.

I would also like to thank colleagues with whom I have written papers or research software, or who have otherwise encouraged and inspired me in this new phase of my career. I wish to especially thank: Sean Batt, Bruce Bimber, Noshir Contractor, Bill Dutton, Rachel Gibson, Mathieu O'Neil, Rob Procter, Jamsheed Shorish and Jonathan Zhu.

Finally, I want to thank Kazuko for the constant encouragement and gentle nudging that helped me to finish this project.

1

Introduction

This chapter aims to provide a context for web social science, introducing some of the major themes that are addressed elsewhere in the book.

Section 1.1 provides an introduction to the key technologies and governance structures that underlie the Internet and the web, and presents a timeline of key events (from the perspective of web social science) in the history of the web. Section 1.2 introduces examples of online computer-mediated interaction which feature throughout the book. Section 1.3 introduces three important phases in the conceptualisation of the web: cyberspace, virtual communities and online social networks. Section 1.4 outlines four disciplinary approaches for conducting empirical research using data from the web. Section 1.5 introduces the concept of construct validity in the context of web data. Finally, Section 1.6 looks at whether the web can should be viewed as a tool that people use for achieving social, political and economic outcomes, rather than a force that shapes behaviour.

1.1 THE WEB: TECHNOLOGY, HISTORY AND GOVERNANCE

The starting point for a book on web social science is necessarily a brief introduction to the technology that underlies the web. While the average social scientist will not need to know much about the technology of the web, it is important to know, for example, that the web and the Internet are not synonymous. The Internet came before the web, and the web is in fact built on top of the Internet.

The Internet is a massive, distributed network of computers, originally developed in the US in the 1960s with funding from the Defense Advanced Research Projects Agency (DARPA). Data that is transferred between computers on the Internet is split into relatively small blocks ('packets') which are then reconstituted at the final destination. Packets follow the most efficient pathway to the final destination; if a particular computer is not available they are automatically rerouted. This enables efficient transfer of data and also means that packets can be delivered even if parts of the network are not functioning (the original interest of DARPA was in ensuring communications in the event of war).

For the packets to be successfully sent and received there need to be rules or *protocols* – two critical protocols are the Transmission Control Protocol (TCP) and the Internet Protocol (IP), jointly referred to as TCP/IP.[1] But TCP/IP are not the only important protocols. The delivery of email involves an additional protocol called the Simple Mail Transfer Protocol (SMTP). The *World Wide Web* (or *web*) is a massive distributed network of *resources* – documents, sounds, images (Box 1.1). The protocol that underlies the web is the HyperText Transfer Protocol (HTTP), which allows the development of web pages written in the HyperText Markup Language (HTML) coding language; these are used to access information on the web. The web is therefore built on top of the Internet. While the Internet is a network of computers connected by cables, the web is a network of documents connected by hypertext links.

The word *network* is very important – a major aim of this book is to show that web social science is network-based social science. However, the networks that are discussed in the book are not networks of computers or documents, but networks of individuals, groups and organisations. That is, the web allows individuals, groups and organisations to form and maintain networks and, in doing so, create digital trace data that can be studied by social scientists. While it is relatively easy to conceive of Facebook as a network of individuals, this book shows that other web applications also facilitate networked behaviour.

The web, which is regarded by some as being the 'largest human information construct in history',[2] was invented by Tim Berners-Lee while based at CERN and was publicly released in 1991. Box 1.2 presents a list of important milestones in the development of the web. The focus is on events that are important for web social science, and references to relevant chapters and sections in the book are provided.

The web is commonly understood to have had three overlapping phases of development or eras: Web 1.0, Web 2.0 and Web 3.0 (Box 1.3). Under Web 1.0, webmasters create content that is then read or consumed by users. Web 1.0 websites are sometimes referred to as comprising the *Static Web* since they typically do not allow a lot of interactivity and the information

---

[1]An *internet* (lower-case 'i') is any set of computer networks connected by TCP/IP. The *Internet* (upper-case 'i') is the largest set of networks – this is the open and public set of computer networks that we all use. An internet within a single organisation is called an *intranet* (although, technically, this refers to a set of networks that could be using any protocol, not necessarily TCP/IP).

[2]http://webscience.org/webscience.html

presented (often reflecting organisational goals, products, services) does not change regularly (relative to the constant flow of change on sites such as Facebook and Twitter).

Web 2.0 blurs the distinction between webmasters and users, with blogging tools, social network sites (e.g. Facebook) and microblog services (e.g. Twitter) enabling non-technical people to both produce and consume content. The act of a person both consuming and producing web content has been referred to as 'prosumption' (e.g. Ritzer and Jurgenson, 2010) and 'produsage' (e.g. Bruns, 2008).

BOX 1.1   RESOURCES ON THE WEB

So how are resources on the web found? Resources such as websites are identified via unique numeric IP addresses that consist of four numbers (between 0 and 255) separated by dots. The Domain Name System (DNS) translates an easier-to-remember, character-based, *fully qualified domain name* (also known as the *hostname*, *sitename* or *subdomain*), which is the unique name by which a computer is known on a network, into an IP address.

The hostname comprises two parts (joined by a '.'): the name of the *host* (this is the computer that is connected to the network) and the *domain name*. A domain name usually consists of two parts. A *top-level domain* (TLD) identifies the type of organisation. There are two types of TLD: generic TLDs (e.g. '.com', '.edu') and country-code TLDs (e.g. '.au', '.uk'). A *second-level domain* such as 'google' or 'yahoo' identifies the organisation.

For example, the hostname voson.anu.edu.au consists of the host 'voson' and the domain name 'anu.edu.au', and currently translates (via DNS) into the IP address 150.203.224.58. The generic TLD is '.edu', the country-code TLD is '.au', and the second-level domain is 'anu'.

A *uniform resource locator* (URL) is an address that defines a route to a file on an Internet server (e.g. web server, FTP server). The first part of the address is the protocol identifier, while the second part is the resource name, with the first and second parts being separated by '://'. Thus, the URL http://voson.anu.edu.au/index.html consists of the protocol identifier 'http' indicating that this is a resource that is hosted on a web server, and thus requires HTTP to access it, and the resource name is 'voson.anu.edu.au/index.html'. The resource name is composed of the hostname ('voson.anu.edu.au'), the directory path to the file ('/'), and the file ('index.html').

A *subsite* is a collection of pages within a particular website. For example, the subsite http://voson.anu.edu.au/news is a part of the VOSON project website and contains pages with details on project activities, e.g. http://voson.anu.edu.au/news/2012, http://voson.anu.edu.au/news/2011.

BOX 1.2   WEB TIMELINE

1983 – TCP/IP implemented

1984 – William Gibson publishes *Neuromancer* (Section 1.3.1)

1985 – Domain Name System (DNS) introduced

1990 – The Internet comprises over 100,000 hosts

1991 – Linus Torvalds begins work on the open source Linux operating system (based on the MINIX variant of the Unix operating system) (Section 9.1.1)

1990–1994 – New content-publishing services released, e.g. news/bulletin boards, FTP, gopher (menu-driven system for accessing files), first content search engines (e.g. Brewster Kahle's Wide Area Information Service, WAIS)

1991 – Tim Berners-Lee's World Wide Web is publicly released. The web eventually swamped all other content publishing services

1994 – Netscape web browser released

1997 – Internet Archive starts archiving the web, currently available via the Wayback Machine (Section 4.3.2)

1998 – Sergey Brin and Larry Page publish (and patent) their 'PageRank' search algorithm, paving the way for Google (Section 7.1.1)

Mid-2003 – There are an estimated 180 million registered hosts on the Internet, 40 million websites and between 600 and 700 million users

2003 – Linden Labs launch Second Life virtual world (Section 9.3.2)

2004 – Mark Zuckerberg founds Facebook.com, heralding the rise of social network sites (Sections 3.3.3, 5.1.2)

2004 – Political bloggers play prominent role in US Presidential election (Section 7.3)

2005 – YouTube video-sharing website launched

2006 – Twitter microblogging service launched (Section 5.2.2)

2007 – iPhone launched by Apple, igniting the market for smartphones (Section 1.3.1)

2011 – Social media play prominent role in the Arab Spring and the Occupy Movement (Section 8.2)

BOX 1.3   PHASES IN THE EVOLUTION OF THE WEB

Web 1.0:   *Static Web.* Key languages/protocols: HTML, HTTP. Key applications: websites (hosted by web server software such as Apache), web browsers (e.g. Firefox).

Web 2.0:   *Collaborative Web.* Key languages/protocols: AJAX, RSS, SOAP, XML. Key applications: web blogs, social network services, microblogs, smartphone operating systems (e.g. Android), software as a service (e.g. Google Docs).

Web 3.0:   *Semantic Web.* Key languages/protocols: RDF, SWRL, SPARQL. Key applications: semantic databases, intelligent personal agents.

Web 3.0, or the Semantic Web, involves technologies that make the web more machine-readable, leading to a 'web of data', which is an evolution of the Web 1.0 'web of documents' (Shadbolt et al., 2006). While the technologies underlying the Semantic Web are proven, there is yet to be a general take-up of Web 3.0. While it is possible to retrofit existing websites to make them Web 3.0 compatible, this would entail a massive amount of work, so webmasters are unlikely to do this until there are clear benefits or reasons to do so. The exception is the government sector, where Open Data initiatives are drawing on Web 3.0 technologies. But for the vast majority of the web, while Web 1.0 and Web 2.0 are ubiquitous, Web 3.0 is still in its infancy.

A common feature of all three phases of the web is the use of technologies to help people find the content they want. With Web 1.0, and to a lesser extent Web 2.0, the core enabling technology are the hyperlink, which enables users to efficiently move around the web ('web surfing'), and search engines that index web content and present search results to users. In contrast, Web 3.0 envisages intelligent personal agents finding content on behalf of users by drawing on users' preferences and browsing habits.

Governance of the Internet occurs at two levels: architecture and operation.[3] In relation to architecture, design and refinement of protocol specifications is undertaken by various working groups coordinated by the Internet Engineering Task Force (IETF). Other organisations take specific roles in particular areas. For example, issues relating to transmission media are handled by the Institute of Electrical and Electronic Engineers (IEEE) and the International Telecommunications Union (ITU), and protocols to do with the web are the province of the World Wide Web Consortium (W3C) industry association. The main organisation involved with Internet operation governance is the Internet Corporation for Assigned Names and Numbers (ICANN), which coordinates the DNS, IP addresses and the generic and country code TLD system.

1.2 EXAMPLES OF ONLINE COMPUTER-MEDIATED INTERACTION

This section aims to familiarise readers with several forms of online computer-mediated interaction. The list is not complete, with a focus on the types of online interaction that are discussed elsewhere in this book.

---

[3]While Internet governance is not a focus of this book, governance structures are looked at in Section 9.1.1 in the context of peer production. Also see the discussion on Internet censorship in Section 8.2.2.

Threaded conversations: newsgroups, discussion groups and chat rooms

Newsgroups are repositories of emails set up for different topics, often hosted on the Usenet system (an example is rec.pets.cats – a Usenet newsgroup dedicated to discussing pet cats). *Threaded conversations* occur within newsgroups when individuals make posts to newsgroups (thus starting a 'thread'), and respond to the posts of other people. Discussion groups (or chat rooms) are hosted on the web and are often functionally similar to newsgroups (which do not necessarily involve web technologies). They can be moderated or unmoderated. An example is the chat rooms that are hosted on America Online (AOL). Another example is Slashdot – a popular web-based technology-related forum, with articles and comments from readers. Slashdot has developed its own subculture involving the accumulation of 'karma' scores, with volunteer moderators being selected from those with high scores. Threaded conversations are looked at in Sections 3.3.2 and 9.1.2.

Web 1.0 websites

A static website is the 'face' of Web 1.0. These generally represent organisational web presence, rather than the web presence of an individual person, and they often do not allow for readers of the website to interact with the website authors.[4] Web 1.0 websites are looked at in Chapters 4 and 6, and in Sections 7.1, 8.1 and 9.2.

Blogs

A *weblog*, or *blog*, is a chronologically updated website, typically written by a single author and designed to provide regular commentary on particular topics or else to serve as an online diary. Technically, there is no difference between a static website and a blog: the differences are in how the site is used. However, an innovation that was developed in the context of blogs is RSS feeds which allow blog subscribers to know when new content has been posted. Blogs are looked at in Section 7.3.

Wikis

A wiki is a website where web pages can be edited by members of the public, using a simplified markup language. Wikis are designed to enable non-technical people to jointly collaborate on the creation of web content, with the best-known example of a wiki being Wikipedia. Wikis are looked at in Section 9.1.2.

---

[4]Note that organisational websites are increasingly incorporating Web 2.0 features (e.g. blogs and RSS feeds), so the boundaries between Web 1.0 and Web 2.0 are blurring.

Social network sites

Social network sites are websites that allow people to create personal profiles and interact with other people with profiles by requesting and accepting 'friendships' and joint membership in groups (representing people who graduated from a particular university, for example, or who share interests). The best-known example of a social network site is Facebook, but notable predecessors were Friendster (in the US) and Cyworld (in Korea). Other examples of social network sites are LinkedIn (for professional networking) and Renren in China. Social network sites are looked at in Sections 3.3.3 and 5.1.2.

Microblog sites

A microblog allows subscribers to broadcast short messages (e.g. a maximum of 140 characters) to other subscribers of the service. The best-known microblog is Twitter, and Sina Weibo is a prominent example of a Chinese microblog ("weibo"). Microblogs are looked at in Sections 3.3.4 and 5.2.2.

Virtual worlds

Virtual worlds are simulated environments where individuals can assume digital representations (avatars) and interact with other individuals. There are two types of virtual worlds. Massive multiplayer online role-playing games (MMORPGs) are typically fantasy-themed and are derived from earlier 'pencil and paper' role-playing games such as Dungeons and Dragons and the first (entirely text-based) virtual worlds, multi-user dungeons. Individuals assume characters (e.g. human, elf), go on quests with other players (e.g. fight monsters and get treasure), use treasure to buy equipment (e.g. armour, weapons) and gain 'experience points' giving the character greater skills. Examples are EverQuest (EQ), published by Sony Online Entertainment, and World of Warcraft (WoW), published by Blizzard Entertainment. The second type of virtual world is exemplified by Linden Lab's Second Life, which is a popular non-gaming virtual world. In Second Life people can build alternative realities online and, while there are rules governing how you construct your avatar and buildings and how you interact with other people, unlike MMORPGs, Second Life inhabitants are not playing a game. Virtual worlds are looked at in Sections 8.3 and 9.3.2.

1.3 CYBERSPACE, VIRTUAL COMMUNITIES AND ONLINE SOCIAL NETWORKS

This book aims to show how empirical social science can provide insights into the impact of the web on society, and how web data can be used to answer long-standing social science research questions. But the Internet is just infra-structure, and the protocols and services that underpin the web are just tools.

In order to achieve the objectives of the book, we need to go beyond the technology and look at how the web is being used by people and organisations, what type of behaviour is occurring on the web, and how this might reflect real-world behaviour and potentially have real-world impacts.

A starting point is a review of three important phases in the conceptualisation of the web: cyberspace, virtual communities and online social networks. As with the technological phases of the web outlined in Box 1.3, the conceptual phases of the web are overlapping.

1.3.1 Cyberspace

The term *cyberspace* was conceived by the science fiction author William Gibson:

> Cyberspace. A consensual hallucination experienced daily by billions of legitimate operators, in every nation, by children being taught mathematical concepts ... A graphic representation of data abstracted from the banks of every computer in the human system. Unthinkable complexity. Lines of light ranged in the nonspace of the mind, clusters and constellations of data. Like city lights, receding. (William Gibson, *Neuromancer*, 1984, p. 69)

Cyberspace has become a *de facto* synonym for the Internet, and has been hugely influential – especially with academics and activists – as a way of describing the Internet as a virtual place in which people interact[5]. Thus, websites are metaphorically said to exist 'in cyberspace' and any interactions between people would similarly be occurring in cyberspace, rather than in the countries where the participants or website servers are located.

Mapping cyberspace

William Gibson's original concept of cyberspace was a visual one, and there is a huge body of work focused on ways of visualising or mapping cyberspace – see Dodge and Kitchin (2000) for an early compilation. Many of the earlier attempts at mapping cyberspace, while technically impressive and visually striking, did not provide much insight for social scientists since they were large-scale maps of the Internet infrastructure.

With the advent of Web 2.0, it has become increasingly common to see academic research featuring maps of the connections between people (e.g. bloggers, Twitter users, Facebook users) and such maps have the potential to be powerful and evocative displays of social processes. For example, the Divided They Blog image (Adamic and Glance, 2005) which is discussed further in Section 7.3 is a powerful visualisation of political homophily.

However, such visualisations are really just a first step in empirical social science research. They are useful for capturing attention and explaining data

[5]http://en.wikipedia.org/wiki/cyberspace

structure, but need to be supplemented with quantitative empirical network techniques such as exponential random graph models (Section 3.2.2). Also, there has perhaps been too much focus on visualisation at the expense of development of appropriate theoretical frameworks. As noted by Janetsko (2009, p. 170), in many cases 'work centering around nonreactive [online] techniques more or less exclusively addresses visualization of phenomena that are perhaps not properly understood'.

The cyberspace ethos

A *cyberspace ethos* developed during the pioneering years of the Internet, and has been influential in forming attitudes and behaviour in relation to inter-actions on the Internet. An aim of this book is to provide a framework for understanding whether this ethos exists today and how it complements or conflicts with other norms, laws and institutions for the real world.

Clarke (2004) identified several aspects of the cyberspace ethos:

- *Interpersonal communications.* Cyberspace is viewed as being for interper-sonal interactions, with organisations having the roles as providers of resources or services rather than participants. Interpersonal communica-tions on the web were greatly enhanced with the advent of Web 2.0, and consequently a lot of web social science is about individual behaviour on the web. However, social scientific web research has also focused on organisational behaviour on the web (see Chapter 6).
- *Internationalism and universalism.* Although the Internet was developed in the US, content and connectivity are technically available to anyone – there are no borders in cyberspace. However, social science research has focused on the digital divide which was traditionally about borders or boundaries preventing equal access to the web (e.g. DiMaggio et al., 2001). But even if everyone has equal access to the web (and equal skills, so no 'hidden digital divide'), while web content might be equally retrievable, it is not equally visible (because of the role of search engines) and this is looked at in Section 7.1.
- *Egalitarianism.* While participants might have particular roles such as mod-erators on lists, there is no hierarchy of authority on the Internet, and people behave as though they are, by and large, equal. But authority and hierarchy do play out on the Internet (e.g. O'Neil, 2009). Also the net-work structure of the web does mean that actors have unequal network positions and hence may experience different behaviour and outcomes. This is looked at later in the context of online collective identity (Section 6.3), reconfiguring access to academic information (Section 9.2.2), and structural holes in Second Life (Section 9.3.2).
- *Openness.* The Internet's fundamental protocols and standards are open to anyone. However, authors such as Zittrain (2008) have argued that the openness of the Internet is under threat with proprietary services such as Apple's iPhone and 'walled gardens' that prevent people moving across social network sites.

- *Communitarianism and mutual service.* Many participants feel that they belong to a community – they both contribute to this community and draw from it. To what extent is the concept of community useful for understanding behaviour on the Internet (Section 1.3.2)? Economic research into open source communities suggest that, rather than altruism, participants may expect deferred benefits (e.g. labour market reputation) – see Section 9.1.2.

- *Freedoms.* A core aspect of the cyberspace ethos is the importance of personal freedom in cyberspace. In fact, many believe that there should be greater personal freedom in cyberspace than in the real world, and they resent activities by governments and corporations to constrain freedom (e.g. censorship and copyright). This has been famously captured in John Perry Barlow's 1996 *A Declaration of the Independence of Cyberspace*:

  Governments of the Industrial World, you weary giants of flesh and steel, I come from Cyberspace, the new home of Mind. ... I declare the global social space we are building to be naturally independent of the tyrannies you seek to impose on us. You have no moral right to rule us nor do you possess any methods of enforcement we have true reason to fear.[6]

- But to what extent is it reasonable to expect to have more freedom on the Internet than in the 'real world'? How does this aspect of the cyberspace ethos come into conflict with a core function of government: authority (Section 8.2)?

1.3.2 Virtual communities

While cyberspace is an evocative and influential concept, it does not help us to establish a framework for quantitative analysis of online behaviour. From a research perspective, we are interested in being able to operationalise the concept of cyberspace as a virtual place in which people interact. The most common term (other than cyberspace) used to describe this virtual place where people interact is *virtual community*, which was introduced by Rheingold (1993) and is now seen as being analogous to *online community*. In this section, we look at the concept of virtual community and see how it relates to the concept of community as developed in sociology.

Membership of *social groups* or categories can be defined on the basis of personal or individual characteristics such as ethnicity or sex. Social groups are therefore objectively defined: you are either in or out of the group, and group membership does not necessarily involve interpersonal relations.

---

[6]https://projects.eff.org/~barlow/Declaration-Final.html

When can we call a group of people a *community*? For community to exist there needs to be a sense of group attachment and belonging – a shared sense of 'one-ness' or 'we-ness' that can be referred to as *collective identity*. So one definition of community is that it is a group of actors who share a collective identity. But the definition is circular, because at the same time a community can be viewed as a vehicle for the emergence of collective identity, via providing common beliefs, norms and shared understandings (Durkheim, 1964).

How does a group of people develop common beliefs, norms and shared understandings? Three important factors have been identified. First, there is a high degree of perceived homogeneity among members on important criteria such as ethnicity and religion (Gusfield, 1975). Second, there is physical proximity or co-location of individuals (e.g. in villages or neighbourhoods). Third, there is the existence of social relations or ties between actors. Taylor (1982) argues that there need to be direct, multiplex and durable relations that are governed by reciprocity and strong interdependence between members. Barry Wellman defines a community as 'networks of interpersonal ties that provide sociability, support, information, a sense of belonging, and social identity' (Wellman, 2001, p. 228). Wellman's view on community is particularly relevant here, with his emphasis on networks of interpersonal ties and no mention of physical co-location of actors as a pre-condition for community. As put by Wellman, 'I do not limit my thinking about community to neighbourhoods and villages' (p. 228).

It is important to emphasise that sociologists have traditionally regarded shared interests as not being sufficient for the existence of community. Durkheim (1964) argued that shared interests are not enough, and there need to be ties based on emotions, while Weber (1922) emphasised the need for feelings of group attachment. However, there is the more recent concept of *community of interest*, where all that the members share is a common interest – they do not necessarily exhibit emotional attachment to or form social ties with other members of the community.

What is the definition of an *online group*, and when can it be called an *online community* or *virtual community*? How does the above definition of community translate into the online world?

An *online group* can be defined as a group of people who conduct personal computer-mediated interactions, where interaction is focused on a topic that reflects the common interests of the group. Drawing on the above discussion on community, an online group is therefore a group of people with shared interests who communicate via the Internet, but where collective identity does not exist. It should be noted that others use the term online or virtual community for what we are calling here 'online group' (the above definition in fact draws from the definition of online community used by Matzat (2004b)). However, we use the term 'online group' here as it more clearly indicates the absence of collective identity. An

Introduction

11

online group can therefore equivalently be referred to as a *virtual community of interest*.

A *virtual community* is an online group where there are additionally shared values, norms and understandings. What is the role of the three factors identified above as being important for the formation of collective identity (and hence community) – homogeneity, proximity and social ties – in the formation of virtual community? The role of homogeneity (on the basis of characteristics such as race, religion and ethnicity) is surely diminished since people can interact online without revealing much about these personal characteristics. The importance of physical proximity has also been greatly reduced: the Internet allows people located anywhere in the world to connect.

That leaves us with social ties, and at first glance one might argue that this factor alone has retained its importance in the formation of online communities, and has possibly even gained importance (given the potential diminishing of the roles of homogeneity and proximity). Barry Wellman famously declared that 'a computer network is a social network' (Wellman, 2001, p. 227). Rheingold (1993, p. 5) defines an online community as a group of people who hold computer-mediated discussions on a topic for a sufficiently long time with sufficient emotional involvement, and who form relationships: 'Virtual communities are social aggregations that emerge from the Net when enough people carry on those public discussions long enough, with sufficient human feeling, to form webs of personal relationships in cyberspace.'

But it should be finally noted that a virtual community might exist even in the absence of homogeneity, proximity and social ties if the topic of interest that draws the community together is one that itself is value-driven, that is, the members would not be interested in the topic if they did not share common values. This is best explained using an example. The rec.pets.cats newsgroup (where people discuss their cats, i.e. how to care for them) is a good example of an online group while alt.non.racism (a newsgroup devoted to discussing racism, presumably from the point of view that it is morally wrong and should not be present in modern society) is an example of a virtual community.

1.3.3 Online social networks

With the rise of Facebook and other social media, the term 'online social networks' has become increasingly popular. In his revised book on the virtual community, Rheingold (2000) states that had he read work by Barry Wellman earlier, he would have used the term 'online social network' instead of 'virtual community'.

The formal definition of an online social network is covered in Chapter 3, but we note here that a distinction can be made between the terms 'social network site' and 'online social network'. The former refers to an online environment such as Facebook, while the latter refers to the formal representation of a social network, where the data on ties and nodes are the result of online interactions between individuals (perhaps in a social network site such as Facebook).

Similarly, it is important to emphasise that the terms 'virtual community' and 'online social network' are not synonymous. On the one hand, it is possible to conceive that every virtual community can be represented as an online social network. As Wellman (2001) put it: 'Although not every network is a community – unless you think of NATO or interlocking corporate structures as communities – every interpersonal community is a network.' Thus, using the definition of 'virtual community' above, we would define a newsgroup focused on preventing racism as a virtual community and it would be possible to represent this as an online social network since we could collect the data to represent it as a threaded conversation network (Section 3.3.2).

But it is not the case that an online social network will necessarily be a virtual community. For example, if we extracted a network of real-world friends from their Facebook profiles, we could represent and analyse the data as an online social network (Section 3.3.3). But this would not be an example of a virtual community since it is not the case that these people necessarily share common values and norms leading to collective identity.

Finally, as discussed further in Chapter 3, the term *online social network*, while less ambiguous than 'virtual community', is still not without its difficulties in terms of definitions. In particular, we need to distinguish between the types of connections that exist between participants and whether these are likely to lead to the interdependencies between people that are the hallmark of social networks.

1.4 DISCIPLINARY APPROACHES TO RESEARCHING THE WEB

This section outlines four major disciplinary approaches to conducting empirical research into the web: network science (as practised by applied physicists and computer scientists), network science (as practised by social scientists), information science and media studies. The aim is to give a brief introduction to the various approaches, and then indicate where they are covered in more detail elsewhere in the book. It should be emphasised that we focus here on what sets the disciplinary approaches apart rather than identifying what they have in common. But it should be noted that the boundaries between these approaches are not 'hard' (there is active cross-over), and they might in fact be contested by people working in these areas.

Network science (applied physics and computer science)

We distinguish two variants of network science. The first is that practised by applied physicists and computer scientists who study large-scale networks, with the aim of: (1) measuring the properties of these networks (generating 'stylised facts' or 'empirical regularities') and (2) using statistical-mechanical models to generate simulated networks exhibiting the properties that are observed in real networks.

For example, Barabási and Albert (1999) observed the existence of *power laws* in large-scale networks – many network participants have few or no connections, while a handful are very connected. They explained the emergence of power laws using the concept of *preferential attachment*: in a growing network, new entrants to the network prefer to connect with network participants that are already well connected, thus leading to a 'rich-get-richer' phenomenon. See Section 7.1.1 for further details.

Information science

Webometrics (also known as 'cybermetrics') is an approach for analysing hyperlink data and website usage patterns, drawing on bibliometrics and informetrics (which are subfields of information science). See, for example, Almind and Ingwersen (1997), Björneborn and Ingwersen (2004) and Thelwall et al. (2005).

As discussed in Section 9.2.1, webometrics often involves the use of statistical techniques in an attempt to identify what characteristics of a website and of the people who run the website lead to the acquisition of hyperlinks. In a recent example of webometric research, Barjak and Thelwall (2008) analysed counts of inbound hyperlinks to the websites of life science research teams in order to assess the role of hyperlinks as science and technology output indicators.

Media studies

Media studies is an academic field that draws on both social science and humanities, and is concerned with media content and impact (with a particular focus on mass media). While the term 'network' may be used in a metaphorical way, the media-studies perspective on the web is often characterised by an absence of formal network techniques. Researchers from media studies are often more focused (compared with other social scientists) on the Internet as a transformative technology, that is, the creation of 'citizen journalists' (e.g. Flew, 2007; Goode, 2009) who are challenging old media and transforming (or in some cases, creating) democracy across the globe.

While media-studies research into the web typically does not use formal network techniques, the concept of *issue networks* (e.g. Rogers, 2010a) has been developed as a way of understanding how individuals and organisations

Web Social Science Methods

are using the web to engage with particular issues. However, Rogers (2010b, p. 8) points out that 'issue networks may be distinguished from popular understandings of networks, and social networking, in that the individuals or organizations in the network neither need be on the same side of an issue, nor be acquainted with each other (or desire acquaintance)'. See Section 4.2.2 for further details.

Network science (social science)

The second variant of network science is the one practised by social scientists. Note that social science is being defined here as including sociology, political science and economics. This is a narrow definition as many consider media studies to be a social science and some information scientists regard themselves as social scientists.

How does the social science approach to studying the web differ from the other disciplines? First, compared with applied physicists and computer scientists, social scientists are more concerned about using models of behaviour that are clearly grounded in social science. While the preferential attachment model of Barabási and Albert (1999) generates networks that exhibit the power laws that have been found in large-scale networks such as the web, it is a statistical-mechanical model and the actors or agents in the model do not exhibit behaviour that is realistic from the perspective of a social scientist.

Second, compared with researchers from media studies, social scientists are more focused on how the Internet is used by actors (people, organisations, governments) to pursue social, economic and political ends rather than the Internet as a force that is controlling or changing people's behaviour.

Finally, researchers from information science use webometric techniques which allow the finding of answers to the question 'What are the qualities of the actors receiving the most hyperlinks?', while a more social scientific approach to studying hyperlinking behaviour involves the use of exponential random graph models which can answer the question 'Why do actors make or receive a hyperlink?' (Section 4.2.3).

Areas of cross-over

There are of course examples where there is disciplinary cross-over in web research. For example, in the context of studying the visibility of various political messages on the web, Hindman et al. (2003) identified the existence of power laws in the distribution of inbound hyperlinks to web pages containing political content. This is therefore an example of applied physics being used in political science (Section 7.1.1). Similarly, Escher et al. (2006) used techniques from webometrics and network science to study how the web has changed government nodality (the property of being at the centre of social and information networks) – see Section 8.1.

1.5 CONSTRUCT VALIDITY OF WEB DATA

Many concepts in social science are subjective, and it is sometimes difficult to know whether a variable is adequately correlated with the phenomenon of interest that it purports to measure. For example, while IQ ('intelligence quotient') test scores are widely used as proxies for intelligence, some people challenge the *construct validity* of the IQ test: does it measure what is intended (intelligence) or are other factors (education, socio-economic status, culture) going to influence the score to the extent that it diminishes its use as a measure of intelligence?

The construct validity of web data (in particular, digital trace data) is integral to web social science. If one is not able to either empirically or theoretically demonstrate the construct validity of web data for social science research, then one is left wondering why one should, as a social scientist, care about hyperlinks, tweets, Facebook friendships, etc.

This book shows how the construct validity of web data can be assessed in three ways. First, the construct validity of web data may be assessed by testing whether the online network displays *structural signatures* that are consistent with those displayed by real-world actors. For example: Does Facebook friendship network data display homophily on the basis of race, ethnicity (Section 5.1.2)? Are divisions between different groups in the environmental social movement evident in hyperlink networks (Section 6.3)? And to what extent is political affiliation reflected in political blog networks (Section 7.3)?

Second, it may be possible to assess construct validity by testing whether variables constructed from web data are correlated with other accepted measures of the construct. For example, if counts of inbound hyperlinks to academic project websites are correlated with other characteristics of academic teams (e.g. publications, industry connections) that are used as proxies of academic authority or performance, then this is evidence of the construct validity of hyperlink data in the context of scientometrics (Section 9.2.1). In Section 7.1.1 the construct validity of hyperlink data is assessed in the context of the visibility of political information. The argument is made that counts of inbound hyperlinks are likely to be correlated with numbers of visitors to websites ('eyeballs'), and to the extent that the latter is an accepted measure of political visibility, the former therefore has construct validity.

Finally, the construct validity of web data may be demonstrated if it can be shown that an actor's position in an online network has influence on his or her performance or outcomes in a manner that accords with what is found offline (Sections 5.2.2 and 9.3.2).

1.6 SHAPING FORCE OR SOCIAL TOOL?

The final consideration that helps to provide context for this book is the question of whether the web has changed behaviour or is more a tool that people use to pursue their social, economic and political ends. While there is no doubt that the web has had (and is continuing to have) a remarkable impact on the world, and many researchers focus on understanding how the web transforms behaviour, the focus of this book is more on the latter question.

In considering the impact of the web (and, in particular, the concept of a virtual community), Fischer (1997) drew on his previous research findings that the influence of new technologies on patterns of communication and community was moderate, in comparison to other factors such as demography and economic forces. This led Fischer to conclude that 'we ought to think more about [new technologies] as tools people use to pursue their social ends than as forces that control people's actions' (p. 115).[7]

The present book is not focused so much on the web as a transformative technology but rather as a technology that people make use of in their social, economic and political behaviour. The book is focused on types of behaviour that have been studied by social scientists for a long time, but identifies the opportunities and challenges that are presented by digital social data.

1.7 CONCLUSION

This chapter has provided an introduction to the key technologies that underlie the web and has outlined some of the major events and phases in the development of the web. Prominent examples of online computer-mediated interaction that feature elsewhere in the book were introduced. The chapter also aimed to provide an introduction to web social science, showing how it differs from other academic approaches for studying the web. This was done by first outlining three key phases in how people have conceptualised the web: cyberspace, virtual communities and online social networks. It is the latter approach (online social networks) that is most relevant to this book.

Another way of distinguishing web social science is to look at various disciplinary approaches to studying the web, and this chapter identified four such approaches: network science (as studied by applied physicists and computer scientists), information science, media studies and network science (as studied by social scientists). It was argued that the social science approach to network science is distinct from the other three approaches. Web social science (as presented in the remainder of the book) draws mainly from social scientists' perspective on network science, although contributions from applied physics and information science also feature.

Finally, it was noted that a key distinguishing feature of this book is the perspective that, rather than being a force that is shaping human behaviour,

---

[7]It would be interesting to see whether Fischer's conclusion would be different today, since 1997 was early in the history of the web.

the web can perhaps best be viewed as a tool that people use to achieve social, economic and political outcomes. The web provides social scientists with a unique data source for studying this behaviour, thus providing new insights into long-standing questions in social science.

Further reading

Flew (2008) provides an introduction to Internet law, policy and governance. See Bruns (2008) for more on prosumption and produsage. See Rheingold (2000) for more on virtual communities.

PART I

WEB SOCIAL SCIENCE METHODS

2

Online Research Methods

The aim of this chapter is to provide an introduction to the main types of online social research, and show how the methods that are the focus of this book fit within the scope of online research methods more generally.

Section 2.1 introduces two dimensions of social research: method and researcher presence. It then shows how the four main modes of social research translate to the online world. Section 2.2 introduces online surveys, and online interviews and online focus groups are introduced in Section 2.3. The analysis of digital trace data, a major focus of this book, is then introduced – Section 2.4 introduces web content analysis and Section 2.5 provides a brief introduction to analysing social media data using network techniques.[1] Online experiments are introduced in Section 2.6, and online field research is briefly introduced in Section 2.7. Finally, ethical considerations associated with the use of digital trace data are discussed in Section 2.8.

2.1 DIMENSIONS AND MODES OF ONLINE RESEARCH

In introducing online research methods, it is useful to think of two dimensions of social research: *method* and *researcher presence*.[2] We introduce the method dimension first, and then discuss researcher presence after having outlined the various modes of social research.

---

[1]Social media network analysis is a major focus of this book and therefore has its own chapter (Chapter 3). Chapter 4 is focused on analysis of hyperlink network data.

[2]This chapter refers to approaches traditionally used in social research. More detail on these approaches can be found in social research texts such as Babbie (2007) and Neuman (2006).

There are two main social research *methods*:

- *Quantitative methods* typically involve the use of standardised research instruments for collecting data from a sample which has been drawn from a larger population. Quantitative researchers generally work with numbers. Quantitative methods often focus on a deductive approach, where hypotheses that are generated from a particular theoretical model or framework are tested using statistical techniques.
- *Qualitative methods* are often used to explore concepts. They typically do not involve the use of sampling techniques or standardised research instruments, and qualitative researchers generally work with text. Qualitative analysis methods are generally inductive, where the researcher starts with the observations and identifies patterns that can be generalised to a particular theory.

There are four main *modes* of social research:

- *Experiments* involve an attempt to establish the influence of a particular variable of interest on behaviour by making an explicit comparison between two similar groups of individuals: one group (the treatment group) is exposed to a change in the variable of interest, while the other group (the control group) is not exposed. The impact of the changed condition can then be assessed. There are four main types of experiments:

  — *Laboratory experiments* are conducted in environments that are set up specifically for this purpose, and the researcher thus has a lot of control over the conditions. Individuals are specifically recruited to participate. While laboratory experiments allow the researcher to control for confounding factors that may compromise the ability to accurately assess the impact of the variable of interest, this mode of research is often problematic for social research as it is difficult to accurately and ethically replicate the types of environment that social researchers are interested in. For example, it is difficult to study social influence (Section 5.2) in a laboratory experiment.

  — *Field experiments* are conducted in a natural environment, that is, outside the laboratory. Participants are generally not aware that they are participating in a field experiment. With field experiments the researcher still has control over the variable of interest (and hence which person is exposed or not), but since the experiment is being conducted in the real world, it is much harder to control for confounding factors, compared with laboratory experiments. An early example of a field experiment is the New Jersey Income Maintenance Experiment in which randomly selected low-income families had their tax and benefits arrangements altered in order to assess how this changed work and consumption behaviour (Ross, 1970).

— *Natural experiments* are conducted in the real world (like field experiments) but involve the researcher taking advantage of a naturally occurring change in conditions faced by participants. An example of research involving a natural experiment is Sacerdote (2001), where a residential college policy of randomly assigning roommates in the first year of college was used to assess peer effects in education outcomes. Another example is that of Costello et al. (2003) who studied the impact on child mental health of an exogenous increase in family income associated with the opening of a casino on a Cherokee reservation in rural North Carolina.

— Some authors distinguish natural experiments from *quasi-experiments* (see, for example, Remler and Van Ryzin, 2011, chapter 13). With the former, the treatment or intervention is often planned, but it is not aimed at influencing the outcome that is the focus of the study. The intervention used by Costello et al. (2003) had the goal of raising community living standards, not the improvement of child mental health outcomes, and hence this was a natural experiment. In contrast, with quasi-experiments, the intervention is aimed at influencing the outcome that is being studied but there is not full random assignment of individuals or households, and hence it is not a field experiment. An example of research involving a quasi-experiment is McKee et al. (2007) who studied the impact of a government initiative aimed at encouraging children in Scotland to walk to school. The authors compared children from participating schools with those from schools with similar socio-economic and demographic profiles, but which did not participate in the initiative. Because the children were not randomly assigned between participating and non-participating schools, this is a quasi-experiment rather than a field experiment.

- *Surveys, interviews and focus groups* involve the application of a research instrument such as a questionnaire in the case of a survey, or a set of questions, guidelines or a script for interviews and focus groups. While it is convenient to group these three research modes together, surveys are examples of quantitative research, while interviews and focus groups are used in qualitative research.

- *Field research* involves researchers observing people and events as they naturally occur. There are two basic types of field research:

— *Participant observation* is where the researcher openly identifies him- or herself and interacts with research subjects. An example is Hochschild (1997) who observed employees at a large company as part of research into work–family balance.

— *Non-participant observation* is where subjects are aware they are being observed but have no information on the study's nature. An example is Lofland (1973) who studied strangers interacting in public places.

- *Unobtrusive research* is where the researcher has no contact at all with the research subjects. There are two main types of unobtrusive research:

  — *Secondary data analysis* is the use of data collected earlier by someone else for another purpose. An example would be using census data to study premarital childbearing: the census data were not collected specifically for the purpose of the study, but can be used in that research.

  — *Content analysis* is the study of recorded communications. An example is Pescosolido et al. (1997) who analysed children's books to see how the portrayal of black characters was influenced by prevailing social conditions in the US. Another example of content analysis is the use of diaries to examine social life (e.g. Elliott, 1997; Johnson and Bytheway, 2001). Diaries are increasingly used in social research on topics such as health behaviour, gender and sexuality, daily expenditure, and how people spend and use their time. There are two main types of diaries used for social research: unsolicited (spontaneously maintained by participants) and solicited (created and maintained at the request of the researcher). Diary research therefore uses data that is created by the private behaviour of individuals (behaviour that is not directed to someone else). Of more relevance to this book is research involving the traces of social behaviour (behaviour directed at others) that people create. Before the Internet such data were hard to come by, but they are now created in vast quantities in social media environments such as Twitter and Facebook. Finally, we can further distinguish quantitative and qualitative content analysis, with the former involving the analysis of text using numerical techniques while with the latter the text is not reduced to numbers.

The first three modes of social research listed above (experiments; surveys, interviews and focus groups; field observation) all involve some researcher–subject contact, which can potentially lead to changes in behaviour: this is known as the 'Hawthorne effect' (see Neuman, 2006, p. 265). This problem is avoided with unobtrusive research. So it is useful to classify the research modes according to the dimension of *researcher presence*:

- *Obtrusive or reactive research* – experiments; surveys, interviews and focus groups; field research.
- *Unobtrusive or non-reactive research* – secondary data analysis; content analysis.

The discussion of method and researcher presence can be used to categorise the various modes of online research (Table 2.1). The bottom row of the table is a subset of research modes that involve *digital trace data*: quantitative and qualitative web content analysis, social media network analysis, online field experiments and online natural experiments.

Table 2.1 Modes of online research

Researcher presence	Research method	
	Quantitative	Qualitative
Obtrusive/Reactive	Online surveys Online laboratory experiments	Online interviews Online focus groups Online field research

Unobtrusive/
Non-reactive
	Quantitative web content
	  analysis
	Social media network
	  analysis
	Online field experiments
	Online natural
	  experiments
		Qualitative web content
		  analysis

In the remaining sections, the various online research methods are discussed further. Social media network analysis is a major focus of this book, and Chapter 3 is devoted to it.

2.2 ONLINE SURVEYS

Survey data collection involves standardised questionnaires administered to a sample of the target population.[3] Online surveys are surveys that are delivered via email and the web.[4] We can compare online surveys with traditional surveys that are delivered via postal mail and telephone.

The use of online surveys is increasing. In 2006, according to sources cited in Vehovar and Manfreda (2009), 76% of US research organisations conducted online surveys, and online surveys were the primary mode for 32% of organisations. Internet surveys accounted for 20% of global data collection expenditure in 2006. However, it is still the early days of Internet surveys, and there are some methodological problems, which we discuss in this section. But the cost and efficiency are such that online surveys will increasingly be used, and solutions to methodological problems will hopefully be found.

---

[3] This section draws from Vehovar and Manfreda (2009) and Fricker (2009).

[4] Some authors define online surveys as surveys delivered over any network including local area networks within organisations, mobile phone networks (SMS surveys), etc., and Internet surveys as surveys delivered via email or the web. Thus, according to these authors, Internet surveys are a subset of online surveys. In this book, we ignore this distinction and focus on web/email surveys (we use the terms 'Internet survey' and 'online survey' interchangeably), since the methodological considerations are common across any survey delivered over a network.

2.2.1 Sampling: basics

To help contextualise the later discussion of methodological challenges associated with online surveys, a brief introduction to sampling is provided in this section. More detail about sampling in social research can be found in Babbie (2007) and Neuman (2006).

Sampling can be used to make statistical inferences about the characteristics of a larger population, by collecting data on a subset of the population. Sampling can have several advantages over conducting a *census*, in which every unit in the population is surveyed. First, sampling is cheaper and easier to administer, compared with conducting a census. Second, sampling is more accurate – even though sampling introduces sampling error (see below), this can be more than offset by reduction in non-response error (see below) because more resources can be directed towards survey design, piloting and non-response follow-up.

Sampling error

Sampling error is the difference between the sample and population that is due to the process of sample unit selection. There are two types of sampling error:

- *Chance error* is when unusual units just happen to be drawn into the sample. This can be minimised by using large samples.
- *Sampling bias* is a systematic tendency to survey units with particular characteristics. There are two types of sampling bias:
  - *Coverage bias* is when some part of the population cannot be included in the sample and is thus absent from the *sampling frame* (a list of those within the population who are eligible to be sampled). As an example of coverage bias, assume we want to survey older Australians (this is the target population). We have a sample frame of email addresses of members of a national lobby group for seniors. There are two sources of coverage bias here: to be in the sample a person has to (1) be a member of the lobby group and (2) provide an email address.
  - *Non-response bias* is when sampled units with particular characteristics are less likely to respond. As an example of non-response bias, assume we are wanting to survey internet-using members of a national lobby group for seniors (this is the target population). We have a sampling frame of email addresses provided by the lobby group, but the survey can only be completed using the latest version of the Internet Explorer web browser (and there is no email survey available as an alternative).

Sampling error can be contrasted with *non-sampling error*: the difference between the sample and the population that is unrelated to sample unit selection (it could also occur with a census). An example of non-sampling error is measurement error, where the survey response differs from the true response because of, for example, a faulty questionnaire.

Types of sampling

There are two major types of sampling:

- *Probability-based (random) sampling* is where the probability of a unit being included in the sample is known. Examples are:
  — simple random sampling, where each unit of the population has an equal chance of being selected into the sample;
  — stratified random sampling, where the probability of selection into the sample is related to a characteristic (or characteristics) of the units of the population, for example, geographic location of households.

- *Non-probability sampling* is where the probability of a unit being included in the sample is not known. Examples are:
  — convenience sampling, where units of the population are chosen on the basis of ease of access;
  — snowball sampling, where additional waves of respondents are drawn from friends or contacts of existing respondents.

Generally, only random sampling can be used for statistical inference about a larger population.

2.2.2 Types of Internet surveys

Corresponding to the two types of sampling, there are two major types of Internet surveys: probability-based and non-probability.

Probability-based Internet surveys

There are four types of probability-based Internet surveys:

- *List-based sampling* (web, email). Here, the sample is drawn from a list of population units. This approach is mainly useful for surveying populations for which a complete list of population units is available, for example, a list of email addresses of employees of a particular organisation. This method is unsuitable for surveying the general population, since a complete email list is not available.
- *Non-list-based random sampling* (web, email). The traditional approach here is random digit dialling (RDD). This method cannot be applied to Internet surveys because it is not viable to randomise email addresses.[5]

---

[5]Zhu et al. (2011) have developed an analogue to RDD for conducting probability-based sampling of blogs.

- *Pre-recruited panel survey* (web, email). This involves a group of individuals recruited from the general population participating in a series of surveys over time. If needed, participants are provided with an Internet connection and computer. These surveys are representative of the general population, but there is the problem of *panel conditioning* where, over time, the sample becomes less representative of the general population (but this problem is not unique to online surveys).
- *Intercept or pop-up survey* (web). Visitors to a website are presented with a pop-up at certain intervals (e.g. every 10th visitor). This approach is generalisable to the population of visitors to the site, but a problem is that non-response can be high, and there is no way of controlling for it (since there is no information on those who chose not to respond). There may be bias, for example, towards more satisfied customers, and away from heavy web users who tend to ignore pop-ups. Further, particular types of visitors to the site might have differential use of pop-up blocker software.

Non-probability Internet surveys

There are two types of non-probability Internet surveys:

- Surveys using harvested email lists (web, email), e.g. collected from website postings or when people sign up at websites, and sold on by email brokers. The problem here is that this approach may violate professional ethical standards and possibly even be illegal. Response rates are likely to be very low (since people generally do not like having their email addresses harvested).
- Unrestricted self-selected surveys (web). These are open to the public for anyone to participate in. They are not representative of the general population.

2.2.3 Online surveys: process and ethics

While online surveys provide many options regarding survey features (e.g. different question formats; the ability to randomly allocate participants to different versions of the survey; randomisation of questions; multimedia stimuli), Best and Krueger (2009) warn that different hardware, software and Internet connectivity can lead to unintended changes in the appearance and functioning of the survey instrument. This can impact on the quality or extent of data collected, and for this reason, the authors advise researchers to design their online surveys carefully to ensure that they are delivered to participants in a uniform and usable manner. If possible, researchers are recommended to make use of third-party online software tools and services, and Kaczmirek (2009) provides advice on how to select providers.

Survey research has well-developed ethical guidelines (see the references in Vehovar and Manfreda, 2009), and these are largely applicable to online surveys. However, there are some ethical issues that are specific to online surveys, such as: using unsolicited email to contact participants; obtaining informed consent from participants; privacy and security threats arising from the fact that online survey databases are necessarily connected to the Internet, and hence vulnerable to hacking; and unintended surveying of children. See Enyon et al. (2009) for a discussion of the ethics of Internet research more generally.

2.2.4 Online survey example: election studies and election polls

Jackman (2005) assessed the online polls that were run during the 2004 Australian federal election and found significant *mode effects* (where the survey approach has an effect on the composition of the sample or the way the respondent answers the survey, and hence the analytical results). The author found that three telephone polls run during the election had the lowest bias (the difference between what the survey predicted in terms of voter intentions and how the vote actually turned out on the election day), while there was medium bias for a face-to-face survey, and the highest bias for the online poll.

However, the online poll discussed by Jackman (2005) was an open or unrestricted poll and hence resulted in the collection of a convenience sample. Faas and Schoen (2006) similarly found that an open online poll used in the 2002 German federal election was heavily biased in terms of the percentage of people in various socio-demographic categories and political involvement (interest in politics, likelihood of voting, party affiliation). The sample was also biased in terms of associations amongst variables, with a stronger correlation between urban/rural location and left/right party affiliation than was found with random sampling.

Faas and Schoen (2006) noted that with an open online survey, there is no clear-cut sampling frame. There are three thresholds of participation: the person becomes aware of the survey; the person has access to the Internet; and the person decides to participate. To the extent that online surveys are advertised on the web, heavy Internet users are more likely to encounter them. Advertisements are also more likely to appear on sites with political content, and hence people who are interested in politics are more likely to see them. The self-selection nature of online polls means that they produce results that are significantly different from the other survey modes, since those who are more deeply involved in politics will be over-represented.

Jackman (2005, p. 514) felt that 'it would seem hasty to write off the Internet as a polling medium' but that the 'thorny issue' of the non-representativeness of self-selected samples needs to be resolved. But this will be challenging in the case of Australia for two reasons. First, compulsory voting in Australia means that the target population is the general population. Second, the characteristics predicting Internet usage are correlated with politically relevant characteristics (age, rural/urban location). Jackman (2005) cited work on deriving weighting schemes to make self-selected Internet samples broadly representative of general populations.

Finally, it should be noted that there may be very little bias arising from the use of Internet election surveys when probability-based sampling methods are used. Sanders et al. (2007) estimated multivariate models of voter turnout and party choice using data from the 2005 British Election Study and found few statistically significant differences between model coefficients generated using the data collected via the Internet survey and those from the data collected in person. The authors noted that 'In general, the in-person and Internet data tell very similar stories about what matters for turnout and party preference in Britain' (p. 257).

2.2.5 Other issues

Online surveys tend to have low response rates, 6–15 percentage points lower than other survey modes (Manfreda et al., 2008). In light of this, researchers may be inclined to use the following approaches to increase sample sizes, but it needs to be realised that there is no 'quick fix' for improving response rates for online surveys (see Fricker, 2009; Vehovar and Manfreda, 2009, and references cited therein):

- *Incentives.* The use of incentives has been found to have little effect on response rates for pop-up surveys and web-based surveys.
- *Advanced graphics and multimedia.* Regarding questionnaire design, advanced graphics and multimedia should only be used to help respondents' understanding of the questions. There is no evidence that fancy graphics and multimedia improve response rates.
- *Mixed-mode surveys.* This is where different modes of survey (e.g. Internet, phone or mail) are used in an attempt to improve the representativeness or overall response rate of a sample. However, researchers need to be conscious of the potential for mode effects (discussed previously in the context of online polls). Respondents tend to favour traditional modes (phone or mail), so it is perhaps best to only use traditional modes for following up non-respondents. It also needs to be realised that the use of mixed-mode surveys raises costs, thus working against one of the perceived benefits of online research (lower cost).
- *Sending to the entire sampling frame.* Because of the low marginal cost of Internet surveys, there is a temptation to send them out to the entire sampling frame. But in doing this a probability sample is being forgone for a convenience sample (allowing participants to opt in). It needs to be remembered that it is more important to worry about measurement, coverage and non-response bias, than about sampling error.

2.3 ONLINE INTERVIEWS AND FOCUS GROUPS

The focus of this section is initially on online interviews (a lot of the points carry over to online focus groups) and then on issues that are specific to online focus groups.[6]

2.3.1 Types of online interviews

There are two main types of online interviews: asynchronous and synchronous. An asynchronous online interview is one where the researcher and participant do not need to be engaging in the interview process at the same time, while synchronous online interviews require that they are interacting at the same time (notwithstanding delays involved with packets of information travelling between computers).

Asynchronous online interviews

Asynchronous online interviews can be conducted via email or online forums. However, if an online forum is to be used, one should consider whether to use an existing online forum or create a new one. Additionally, there are privacy issues associated with running an interview via an online forum.

The advantages of asynchronous online interviews are as follows:

- Email is pretty simple and pervasive.
- Interviewees can answer at their own convenience.
- There are no time restrictions (this is particularly helpful when participants are in different time zones).
- Interviewees can take time to consider responses.
- Interview transcripts are automatically created.

However, asynchronous online interviews also have some disadvantages:

- Technical competence may be inhibited by disabilities (e.g. dyslexia, visual impairment). However, there are people with other types of disabilities who can be included in the study if the interview is conducted via email, compared with face-to-face.
- Having time to consider responses may lead to 'socially desirable' responses rather than spontaneity.
- It is easy for a respondent to ignore emails, i.e. drop out of the study.
- It has been argued that the quality of data is not as good as that obtained from face-to-face interviews.

---

[6]This section draws from O'Connor et al. (2009) and Gaiser (2009).

Synchronous online interviews

Synchronous online interviews are conducted, for example, via conferencing software, chat (either client- or server-based), or Voice over Internet Protocol (VoIP; e.g. Skype). There has been relatively low uptake of synchronous interview methods (compared with email interviews), and this is possibly related to technological challenges. Synchronous online interviews have a number of advantages:

- They more closely resemble traditional face-to-face interviews.
- They are more spontaneous – it is harder for people to construct socially desirable responses.
- They enable the running of focus groups (people can interact).
- Automatic transcripts can be generated (if using chat).

The main disadvantage of synchronous online interviews is that the fast-paced nature of the interview means that transcripts can be disjointed and not follow a sequential form.

2.3.2 Online interviews: process and ethics

Recruitment

The Internet can make things simpler since it provides access to people with particular interests (e.g. members of online communities of interest such as people with a particular health problem). But how to get access to the community? A common approach is to contact the discussion board moderator or website provider and ask for permission to contact participants. With permission, you can then post a message to the online forum or email people on the newsgroup. Note that 'netiquette' requires that postings to online forums or emails to newsgroups should be relevant (which the research project may not always be), so there could be ethical issues associated with this.

Representativeness

As noted above, representativeness is less of an issue with qualitative research. There may be biases (e.g. restricted to US-based sites or English-language sites), but is this any different from on-site (real-world) qualitative research? As long as the researcher is clear that research results are not meant to be generalised internationally, this will probably not be a problem. As we saw above with online surveys, representativeness is less of a problem if you are researching a particular group of Internet users.

Conducting the interview

In conducting an online interview, researchers often try to replicate face-to-face methods (the gold standard), and strategies are required to compensate for loss of visual cues.

As with an offline interview, you will generally need to provide an introduction to the interview, stating the aims of the study, explaining the process, etc.

A major problem with online interviews is the difficulty of verifying the true identity of the person being interviewed. Introductory material might include profile data on the interviewer(s) in an effort to indirectly elicit similar socio-demographic data from respondents (that would be visually apparent in a face-to-face interview, e.g. sex, age, race). However, online it is possibly more acceptable to be more direct than in face-to-face situations: users realise that without visual cues, direct questions are needed (e.g. directly asking, 'Are you male or female?').

In terms of building rapport with the interviewee, a university email address may add credibility to the research, but the researcher is more likely to be identified as an 'outsider' to the online community. A web page introducing researchers (e.g. using photos) can help to build rapport, although this process of self-disclosure may actually make the respondent more inhibited.

A final issue is how best to deal with silences – delays in responding to emailed questions. A silence could mean the respondent has withdrawn from the research or they might just have been interrupted. It is not clear how many follow-up emails should be sent before they become intrusive.

Ethics

The ethical considerations that apply to face-to-face interviews carry over to online interviews – is there anything different? It is possibly more difficult to gain informed consent with an online interview. Also, as mentioned above, it is harder to verify participant identity. Withdrawal from a synchronous online interview can be easily facilitated using a 'withdraw' button in the chat window, and hence ethically managing the process of withdrawal is not that different from the offline setting. However, as noted above, managing withdrawal with email (asynchronous) interviews is more difficult, and the researcher needs to be careful to not continue emailing questions or prompts to someone after they have in fact withdrawn from the interview. While it would be very unusual for a participant in a offline interview to simply get up and leave the room, leaving the interviewer wondering whether they had withdrawn from the interview or were coming back, it is possible for the equivalent to happen with email interviews.

2.3.3 Online focus groups

Focus groups enable the researcher to evaluate ideas in a group setting. The group setting provides additional insights from dialogue and interaction between participants. Focus groups are used for public opinion assessment, marketing and social research – they are relatively inexpensive means of collecting qualitative data and testing or developing theories.

The advantages of online focus groups are as follows:

- They are inexpensive.
- They provide access to a broad range of participants (compared with online interviews, where the Internet enables access to specific types of people).
- There is less pressure to conform than in a face-to-face setting.
- The lack of visual cues means participants are less likely to have knowledge about what the moderator thinks about a particular subject. This might mean the group pushes further on a subject initially not interesting to the moderator (but this could be useful).

Risk and vulnerability for participants are different in online focus groups:

- In offline focus groups, participants can be physically identified and their exact contribution to discussion is more obvious.
- In online focus groups, the loss of visual cues may lead to a participant revealing potentially embarrassing information.

Informed consent is important, even though the risks of participating in the research may not be clear. The process of obtaining informed consent allows the researcher to clarify expectations and map out the researcher's obligations to participants, for example, relating to how the technology may affect anonymity, and the risk that someone else may save and use data (e.g. chat logs) that everyone can see.

Researchers will probably need to use asynchronous technologies (email, discussion boards) if participants are in different time zones. A problem is that the moderator will not always be present, resulting in challenges for managing the group.

Synchronous technologies (chat, Skype) are easier for the facilitator or moderator to manage once the focus group session is under way, and are more similar to face-to-face focus groups. However, they can be harder to organise compared to their asynchronous online counterparts since they require everyone to be participating at the same time.

2.3.4 Other issues

The obvious advantage of online interviews and focus groups, compared with their offline counterparts, is they allow researchers to access hard-to-reach populations and expand the geographic boundaries of research at lower cost. However, there is a tendency to judge the quality of online interviews and focus groups by comparing them with their face-to-face counterparts – the latter are seen by some researchers as being the 'gold standard' – and this is not always appropriate.

O'Connor et al. (2009) argue that adapting the face-to-face approach to online interviews is not sufficient – there needs to be online-specific practice. Since online interviews are often supplemented or verified by face-to-face interviews, this suggests it is difficult to use them in a standalone fashion. However, this works against the main advantage of online interviews – expanding the spatial boundaries of research at lower cost.

2.4 WEB CONTENT ANALYSIS

Content analysis is an established analytical approach in the social sciences (see Neuendorf, 2002; Krippendorff, 2004). There are two types of content analysis – qualitative and quantitative – and the differences between the two mirror the differences between quantitative and qualitative research methods described above. Quantitative content analysis involves objectively extracting and analysing content from texts, while qualitative content analysis involves the researcher employing subjective (but scientific) techniques to understand and interpret social reality using texts.

In this introduction to web content analysis, we similarly distinguish between quantitative and qualitative approaches.[7] The definition of *quantitative content analysis* used here is that it involves quantitative analysis of text content extracted from online environments, where the analysis does not involve constructing and analysing networks. This allows us to distinguish *social media network analysis* (see Section 2.5 and Chapter 3) even though it may involve working with text. For example, thread networks (Section 3.3.2) are constructed by parsing the text of forum posts or newsgroups.

2.4.1 Quantitative web content analysis

Quantitative analysis of web content is in many ways similar to quantitative analysis of text content that is not collected from the web. In this section we identify the key features of quantitative content analysis and see how they differ for content collected from the web. To make the discussion more concrete, we compare analysis of traditional *text content* (as an example, we use academic articles) with *web content* extracted from three types of online environment (Web 1.0 websites, social network sites such as Facebook, and microblogs such as Twitter).

A first feature of quantitative content analysis is the unit of analysis (e.g. Riffe et al., 2005). *Sampling units* are what are sampled from the population. In our example, the sampling units in text content analysis are articles from a particular academic field, while with web content analysis the sampling units are websites (e.g. for particular types of organisations), Facebook profiles and Twitter users (or possibly tweets).

Recording units are the items of content that are coded. In our example, with text content the recording unit might be paragraphs or sentences

---

[7]See Herring (2010), who considers whether established methods of content analysis are adequate for use with web content, or whether new methods need to be developed.

extracted from each sampled article. With Web 1.0 content the recording unit might be paragraphs or sentences extracted from the body of the web page, or various meta keyword fields (e.g. title, keywords). A complicating factor with Web 1.0 content is: do you extract recording units from the entire website or just particular parts? For example, in their study of environmental activist websites, Ackland and O'Neil (2011) only extracted meta keywords from the homepages of environmental activist sites, arguing that an organisation should place the most important text (from an organisational positioning perspective) on its homepage. With Facebook the recording unit may be posts on the person's Timeline, for example. With Twitter, the recording unit may be tweets containing a particular hashtag (e.g. all tweets containing the hashtag #climatechange that were tweeted during a particular period).

A second feature of quantitative content analysis is the *sampling approach*. In some cases it might be possible to avoid sampling altogether, and instead conduct a census. For example, in our text content example, it might be possible to collect all articles that have been published on a particular topic in a given time frame. If the population is too large, then sampling approaches need to be employed (e.g. random sampling).

Depending on the research topic, it may be difficult to conduct a census of web content that is to be used in content analysis. For Web 1.0 analysis there is often not an identifiable population of websites to use for a census (or from which to draw a random sample). For example, Ackland and O'Neil (2011) conducted a content analysis with an explicit aim of including in the analysis those environmental activist groups that exist only on the web, in addition to those that have both an online and offline presence such as Friends of the Earth (Section 6.3). In this situation, the population of websites could not be identified in advance and hence snowball sampling was used. If the objective of the study was only to collect data on activist organisations with an offline presence, then it would be possible to sample from, for example, official registers of such organisations.

With Facebook (and other social network sites), it is similarly difficult to identify the population of profiles. For the typical researcher, it is not possible to identify a list of, for example, all profiles belonging to Australians aged 55 years and over. However, it may be that the research is focused on a population of Facebook users that can be identified using some offline criteria. For example, Lewis et al. (2008) collected Facebook data for a cohort of students in a US residential college: this was a clearly defined population of students (Section 5.1.2).

With Twitter, as with Facebook, it is difficult to identify a population of users on the basis of user profiles; that is, it is not possible to identify, say, all Australian Twitter users in a particular category (e.g. using age, sex). However, as noted above, one can use tweet content to isolate populations of Twitter users who are tweeting about particular topics.

A third feature of content analysis is whether the focus is on the *manifest content* (what the author actually wrote) or the *latent content* (what the author meant). Manifest content objectively and unambiguously exists in the text, while latent content is conceptual and cannot be directly observed in the text. A lot of social scientific quantitative web content analysis focuses on the latent content – see the example from Ackland et al. (2010) below.

Examples

Ackland et al. (2010) used quantitative analysis of website content to assess how organisations involved in nanotechnology (manufacturing, research, commercialisation) were engaging with social issues relating to the technology (e.g. public health, risk). The sample of websites was constructed using a snowball sampling approach. Manifest content analysis involved identifying what keywords were prevalent on nanotechnology websites (frequency analysis) and whether these keywords were related to the type of organisation behind the website.

Latent content analysis involved the use of statistical techniques to establish whether website keywords were structured in particular discourses about nanotechnology. Three main clusters or orientations within the debate were discovered. The first, most prominent, dimension was an industrial or proactive discourse which focuses on the prominence of business, investment and opportunity, largely disregarding risk, safety, society, health and, particularly, discussion. The second dimension was a science or education discourse, which strongly emphasised nano-science and featured terms such as industry, science and education, while ignoring topics relating to investment and innovation, as well as health, society and nano-threats. Finally, the third dimension centred on a social or critical discourse, where the health risks associated with the technologies were stressed, along with the need of a political discussion.

A further stage of the analysis involved connecting content analysis with social network analysis. In particular, the authors looked at whether the identified discourses reflect the structure of the network, specifically in relation to the position of different websites within the network. It was found that industrial and technical concerns appear to take precedence over social concerns in the framing of the nanotechnology issue. However, websites aimed at dissemination, particularly blogs and academic sites, do tend to provide a wider perspective on the social and even political implication of nano-technologies.

Hinduja and Patchin (2008) conducted a content analysis of adolescents' Myspace profiles. The research context was the 'moral panic' about the dangers of social network sites (e.g. sexual predators using social network sites to find adolescents who carelessly reveal identifiable information on personal profile pages). The researchers conducted a quantitative content analysis of randomly sampled MySpace profile pages, focusing on manifest content that could be problematic such as the presence of photos where the profile owner or friends were in revealing clothing, or content suggesting that the owner engaged in risky or inappropriate behaviour (e.g. smoking, drinking). The results indicated that the problem of personal information disclosure on MySpace was not widespread, with the overwhelming majority of adolescents using the website in a responsible manner.

2.4.2 Qualitative web content analysis

While qualitative web content analysis is not a focus of this book, we provide an introduction here. Note that we distinguish qualitative content analysis from online field methods (or virtual ethnography), introduced below.

A good way of understanding what qualitative content analysis is (hence its online counterpart, qualitative web content analysis) is to delve further into the comparison between qualitative and quantitative content analysis which was started above. With respect to the issue of sampling units and recording units, and the sampling approach employed, these are still relevant to qualitative content analysis but there is generally less focus on the need for obtaining either a census or random sample of units.

In our discussion of quantitative content analysis we distinguished manifest and latent content. While with quantitative content analysis the focus is often more on the manifest content, with qualitative content analysis the focus is generally more on the latent content. That is not to say that qualitative content analysis does not involve frequency counts of words/phrases extracted from bodies of text, but rather that it will be conducted in a different way compared with quantitative content analysis. As with quantitative content analysis, qualitative content analysis does involve the application of coding schemes; however, the scheme is more likely to be constructed in an inductive/data-driven way, compared with quantitative content analysis.

Examples

Hookway (2008) employed qualitative content analysis of blogs to explore how contemporary urban Australians experience morality in their everyday lives. Blogs are an online equivalent of unsolicited diaries, and Hookway lists particular advantages for qualitative social researchers: publicly available, low-cost and instantaneously available data; naturalistic text data, in the sense that anonymity may mean bloggers are relatively uninhibited; access to populations otherwise geographically or socially removed from the researcher; and the chronological nature of blogs, meaning that they are well suited to facilitating the study of social processes over time.

Hookway (2008) felt that for the topic he was studying, blogs have distinct advantages over interviews (online or offline). The latter rely on the participants' willingness to talk candidly about moral beliefs and actions, and there might be a gap between the participants' 'socially situated subjectivities and their actual practice' (p. 94).

Blogs are private content in the public domain: to what extent might 'performance in blogging' affect the data? Hookway felt that there was a risk that bloggers may strategically frame themselves as having moral or virtuous qualities, but argued that the anonymity of blogging lessens this problem, leading to 'a confessional quality, where a less polished and even uglier self can be verbalized' (p. 97). But the anonymity of blogging can also lead to problems of identity play and deception – how do you know that what a blogger says is true? But Hookway contends that even fabrications provide information about social attitudes/mores and, further, ensuring truth in any research scenario is not straightforward (how do you know someone is being honest in an interview or while filling out a survey?).

Pfeil and Zaphiris (2010) conducted a qualitative content analysis of SeniorNet, an online community for older people, focusing on a discussion board about depression. In terms of the recording unit, the authors noted that while sentences are commonly used in analysis of offline text, it is not always obvious what constitutes a sentence in a discussion board where text is more informal (i.e. conversation). They decided against using the sentence as the unit of analysis, because they felt it would be hard to identify individual sentences, sentences would lead to too many units, and also contextual information that might be evident across several sentences would be lost.

A possibility was to use the message (post) as the unit of analysis, but while it would be easy to segment on this basis, the problem was that posts might contain different topics (even if all related to the overall topic, depression). They decided in the end to use 'the meaning' as the unit of analysis (i.e. what the discussion board authors actually intended). While they noted that this introduced the problem of subjectivity into the segmentation process (since meaning might stretch across several posts by the same author, for example) they provided a set of rules to be applied during the segmentation process, which improved inter-segmentation reliability.

In terms of the development of a set of codes, the authors recommended (given the inductive nature of the coding process they used) that the researcher should first read all the relevant posts to get an impression of the context of the discussion, and once having immersed themselves in this way, should then reread the posts, taking notes and highlighting key terms and themes. They processed 400 messages in this way, using an integrative process to develop the codes and associated descriptions. The messages were then segmented and coded, using a flow diagram, with inter-coder reliability for both segmentation and coding being conducted.

They then used frequency analysis to analyse the data. The authors found that the most frequent category in the dataset was 'self-disclosure' (where the poster discloses personal information). They found that rather than directly asking for support, discussion board participants tended to disclose information about themselves in order to trigger support. The frequencies showed that participants tended to talk about their situation in an emotional way (showing their feeling) rather than in a narrative or medical way.

2.4.3 Web content used in data preparation

Web content is often used in the data preparation phase of research rather than (or in addition to) being analysed in its own right. That is, web content is used to categorise the observations (e.g. websites, Facebook users, Twitter users) so that numeric codes can be applied and quantitative analysis undertaken. For example, Ackland and O'Neil (2011) manually coded environmental activist websites according to the main issues of concern (Section 6.3), and Adamic and Glance (2005) coded political bloggers as either conservative or liberal on the basis of blog content (Section 7.3).

2.5 SOCIAL MEDIA NETWORK ANALYSIS

Many of the types of digital trace data mentioned in Section 1.2 can be represented as networks:

- Data from newsgroups involve threads (postings and responses to postings) that can be converted into networks and analysed as such (Sections 3.3.2 and 9.1.2).
- Hyperlink networks can be constructed by crawling static websites (see Chapter 4).
- Blog networks have been the subject of much research (Section 7.3) and network data can also be derived from Twitter feeds (Sections 3.3.4 and 5.2.2).
- Social network sites generate network data, for example friendships and joint membership of groups (Sections 3.3.3 and 5.1.2).
- Finally, the activities of people in virtual worlds can also be represented as networks, reflecting, for example, friendships in Second Life (Section 9.3.2) or joint participation in guilds or quests in World of Warcraft.

Social media network analysis is a major theme in this book and has its own chapter (Chapter 3).

2.6 ONLINE EXPERIMENTS

Mirroring their offline counterparts, there are three main types of online experiment: online laboratory experiments, online field experiments and online natural experiments.

2.6.1 Online laboratory experiments

As noted above, laboratory experiments are more common in the behavioural sciences and (increasingly) economics than in social research. For a review of advantages and disadvantages of online laboratory experiments for psychological research, see, for example, Reips (2002).

However, online laboratory experiments are increasingly being used for social-science-related research. As discussed in Section 10.2.2, Abbassi et al. (2011) conducted an online experiment to assess the impact of friend recommendations and ratings from the general public on how an individual makes choices. The authors used Mechanical Turk (MT) to recruit participants and conduct the experiment (see also Box 2.1). The advent of MT seems to be at least partly responsible for the increasing popularity and viability of online laboratory experiments for social research.

2.6.2 Online field experiments

Several examples of online field experiments are discussed in this book: Aral and Walker (2010) use two versions of a Facebook application to study social influence and viral product design (Section 10.2.1); Salganik and Watts (2009) construct an 'artificial cultural market' to assess the influence of online rating systems on music choice (Section 10.2.2); and Centola (2010, 2011) uses an artificial social network to study social influence and health behaviour (Section 5.2.2).

2.6.3 Online natural experiments

An opportunity for an online natural experiment in World of Warcraft occurred when a programming error led to an infectious deadly 'disease' (designed to be active only in a small geographic area) spreading to other parts of world, killing many game characters.

Epidemiologists published papers on the 'World of Warcraft Plague' (Balicer, 2007; Lofgren and Fefferman, 2007), but Williams (2010) has questioned the validity of this particular online natural experiment (Section 8.3). Castronova et al. (2009) investigate whether there is a 'mapping' between economic behaviour in virtual worlds and the real world, and make use of a naturally occurring event, the creation of a new version of an existing virtual world (EverQuest 2), to see how quickly economic aggregates in the virtual world (e.g. GDP, inflation) replicate those in existing versions of EverQuest 2 (Section 8.3).

2.7 ONLINE FIELD RESEARCH

The major form of online field research is virtual ethnography, which involves taking the ethnographic tradition of the researcher as an embedded research instrument to online social environments. Offline ethnographic research involves the researcher documenting the language used by community participants and their modes of communication and interaction, in an attempt to understand the shared knowledge of the community. Virtual ethnographic research has been influential in establishing the concept of virtual or online communities (Section 1.3.2) as existing independently of physical space.

BOX 2.1   MECHANICAL TURK

Mechanical Turk is a micro-payment system for crowd-sourcing labour-intensive tasks. MT is an online labour market: 'workers' are hired by 'requesters' to perform tasks. Typical tasks are image tagging, audio transcription, survey completion, and increasingly laboratory experiments. Rewards are as low as $0.01 and rarely exceed $1 – the average hourly rate is about $1.40.

In order to look at the advantages of MT for online experiments Paolacci et al. (2010) surveyed 1000 MT workers in February 2010. Surveyed workers were from over 60 countries (47% US, 34% India). Of the US workers, 65% were female and they had an average age of 36 years (younger than the US population), higher education than average, but lower income than average. Around 14% of US workers said MT was the primary source of income, and for the sample overall, most spend one day or less per week, completing 20–100 tasks.

The authors identified the following the advantages of MT for experiments: the supportive infrastructure (recruiting is generally fast), subject anonymity (which can assist with getting ethics clearance), the possibility of pre-screening workers (e.g. to recruit only women or people who answered a particular way in pre-tests), worker identification numbers (which mean that researchers can re-contact a particular set of workers, i.e. conduct longitudinal research), and the cultural diversity of the worker population. In response to questions about the representativeness of MT workers, the authors found that for US workers, demographics match US population more closely than traditional subject pools (e.g. college students). Further, they argued that non-response error is less of a problem for MT experiments than Internet convenience samples recruited via other means.

Concerning data quality, the question is whether the fact that workers are paid so little means they will not take the experiment seriously. However, there is no evidence that MT data are of lower quality than data from other subject pools. The fact that worker identification numbers must be linked to a credit card prevents workers from providing many separate responses to the same study. All web-based experiments have the drawback that unsupervised subjects tend to be less attentive than if they are in a physical laboratory. This can be controlled for with questions designed to catch inattentive respondents (e.g. 'Have you ever had a fatal heart attack watching TV?').

While virtual ethnography is not a focus of the present book (but see Hine, 2009, for a review), we note that the study by Williams et al. (2006) into group structure and performance in World of Warcraft involved participant observation to 'know what questions to ask [in the online survey], decipher the local language, understand the game mechanics ... [and] the social context of play' (Williams et al., 2006, p. 342). Also, many of the ethical issues for research using digital trace data reviewed in the next section are relevant to virtual ethnography.

2.8 DIGITAL TRACE DATA: ETHICS

This section provides a summary of ethical considerations for using digital trace data.[8] Ethical issues for other online research methods (surveys, interviews, focus groups) were covered above, and they are generally in close accordance with their offline counterparts. Online laboratory experiments should also have similar ethical requirements to their offline counterparts.

Ethical guidelines for use of digital trace data are still a 'moving target', with university ethics committees (known as institutional review boards in the US) still coming to grips with how to assess whether proposed research projects are ethical. Active discussion of these issues can be found on the Association of Internet Researchers (AoIR) mailing list,[9] and AoIR has been involved in the development of ethics guidelines for Internet research (Ess, 2002; Markham and Buchanan, 2012).

The difficulty in establishing what is ethical when researching digital trace data is compounded by the fact that the technology and people's use of the web is changing rapidly. In the era of Web 1.0 and the early days of Web 2.0 it might have been harder to argue that a person maintaining a website or blog was acting in a 'publicly private' manner (see below) since a relatively high level of technological knowledge and intent was required in order for someone to publish on the web. However, the emergence of blogging platforms such as Wordpress and social network sites such as MySpace and Facebook, and the increasing acceptance (amongst younger people in particular) that it is OK to live your life online, have made it much easier for the average person to reveal sensitive personal information, not expecting it would be read by anyone other than a close circle of friends.

The approach typically taken by ethics committees is to use the standards for human-subjects research as the basis for developing guidelines for ethical use of digital trace data. The applicability of the human-subjects model to ethics of Internet research has been questioned in some research contexts, such as where the research involves discourse analysis (see, for example, Herring, 1996), but this model is an adequate basis for most of the research covered in the present book.

There are three major concerns for human-subjects research in social science: informed consent, the distinction between private and public individuals, and participant anonymity. These concerns are now discussed in the context of research using digital trace data.

Informed consent

Informed consent is the process of the researcher informing participants about the nature of the study and any risks that are entailed, and making

---

[8]Note that a lot of these points regarding ethics also apply to online field research.

[9]http://listserv.aoir.org/listinfo.cgi/air-l-aoir.org

sure that this information is understood before obtaining their agreement to participate in the study. There also needs to be a facility for participants to withdraw their consent.

Do researchers of online environments (e.g. websites, newsgroups, blogsites, social network sites, microblogs) need to gain authorial permission when using publicly available content extracted from these environments? Are publicly available online materials such as blog posts academic 'fair game', or is informed consent required? Documents can be researched without approval, but people cannot. Web pages, Facebook profiles and the like are just documents (HTML), so does that mean we can conduct this research without permission?

The key issues regarding informed consent in relation to research using digital trace data are when it is needed and how it can be obtained. This depends on what information is being extracted (and relatedly, what type of online environment it is) and how the information is going to be used. It seems necessary to distinguish between environments that are primarily facilitating information production and sharing (e.g. websites, newsgroups, blogsites, microblogs) and those that are primarily designed for social networking (e.g. social network sites).

With the former, it is perhaps easier to justify the perspective that publicly available archived material on the web is published in the public domain and is therefore fair game in terms of use for research. However, in the situation where, for example, blog posts or newsgroup posts might be reprinted verbatim, there are issues of copyright, and hence 'fair use' needs to be determined, but if being used for academic purposes, then it is unlikely that this would be a violation of fair use.[10] The second perspective is that online postings, while public, are written with expectation of privacy and should be treated as such – hence informed consent is required.

When data are collected retrospectively, there are additional challenges with gaining informed consent since the individuals may have left the newsgroup or have ceased blogging but the data are still available online because the newsgroup continues or the blogsite has been archived (Section 4.3.2). Another issue is that often research will necessarily involve data from all participants in the group or community and it will not be feasible to obtain consent from every single community member.

In summary, with respect to information-sharing platforms, it could be strongly argued that informed consent is not always required. But participant anonymity should be preserved, if possible. Often the question of whether informed consent is required will be dependent on the nature of the community being studied and whether it is a private or public space.

---

[10]Copyright laws have led to some researchers arguing that online text should be used verbatim, and with proper attribution to authors (where identification is possible), but this clearly could not apply in a situation where the nature of the issue being researched requires the identity of research participants to be protected.

With respect to social network sites it is harder to argue against the position that informed consent is required, but again it will probably come down to the public/private question and also anonymity.

Distinction between private and public

In offline contexts, the decision of whether informed consent is required often hinges on whether the data are private or public in nature. However, in the online world, this distinction is blurred. People can reveal information about themselves in a public manner (i.e. blogging) but perceive themselves to be interacting only with a smaller group. So people are undertaking what they think of as private communication in a public space, and have the expectation that anyone 'overhearing' them will not use the information gathered.

But again, it is difficult to know what people are expecting regarding the public nature of their private communications. With newsgroups, a key is the nature of the community and/or the types of people involved. With communities that provide a support role for people with, for example, medical conditions, it may be that the only way the person can get the necessary support is by revealing a certain amount of private information. For example, in their analysis of interactions on a discussion board aimed at seniors dealing with depression, Pfeil and Zaphiris (2010) found that participants tended to disclose information about themselves in order to trigger support.

However, even newsgroups dealing with sensitive topics may still be accessible without informed consent under certain circumstances. For example, Pfeil and Zaphiris (2010) argued that informed consent was not required for their use of the newsgroup data for several reasons: the newsgroup was accessible by anyone and it was not necessary to register to be able to access the material; the organisation hosting the site (SeniorNet) is involved in the education of older people about Internet use and hence it is reasonable to assume that forum participants have a minimum level of technical knowledge, and, in particular, realise that their contribution is public; and finally, there was a tradition of research being conducted on SeniorNet discussion boards and hence participants were likely to have been exposed to researchers previously and become aware of the research outputs using discussion board data.

Herring (1996) contends that the responsibility for the privacy of newsgroup participants lies with both the researcher and the participants (especially those taking an organising role). Participants in an online community such as a newsgroup have a responsibility to clearly declare whether the newsgroup is public or private (and use technology, e.g. membership, to restrict access if the latter). Researchers have a responsibility to respect the privacy of participants (as far as the research objectives will allow), even if the material is publicly available.

With social network sites the distinction between what is public and what is private is clearer because of the use of privacy controls, but researchers need to make a judgement on whether the group being studied is likely to be aware and technically competent enough to exert proper judgement on what they reveal to non-friends (e.g. via adjustment of privacy controls).[11]

Participant anonymity

When should anonymity be granted to research participants? Again, the answer to this question is not clear-cut if the data being collected are publicly available. However, the following guidelines may help researchers to decide when participant anonymity should be protected, even when using publicly available data:

- when a possibly undesired label may be attached to the person or group (e.g. hate group, terrorist);
- when naming the individual/organisation might lead to undesirable effects (e.g. a 'league table' ranking of hate sites, allowing extremist organisations to find out about like-minded organisations with whom to connect);
- when an individual is singled out (e.g. in-depth analysis of an 'average' blogger or Facebook participant).

But preserving anonymity can be very difficult when quoting text since search engines such as Google can be used to search for sites containing the text, thus revealing both the participant and the research site, even if these names are obscured by the researcher.

If the research involves gaining access to lists of friends on social network sites such as Facebook – as in the study of social influence in Facebook by Aral and Walker (2010), discussed in Section 10.2.1 – participant anonymity will most likely be a requirement of gaining informed consent.

Automated data collection tools: other considerations

A lot of the above is written from the perspective that it is at least feasible to contact the individuals or organisations that are authors of web content. But what about situations where web crawlers are used to collect large-scale data, for example the analysis of over 1,500 political bloggers conducted by Adamic and Glance (2005) and discussed in Section 7.3? The scale and nature of the data collected via web crawlers generally make it impossible to contact individuals who manage the websites or blogsites, but there are ethical issues that are specific to the use of web crawlers, as discussed further in Section 4.3.1.

2.9 CONCLUSION

This chapter has provided an introduction to online research methods, overviewing the various modes of online research, and showing how these

---

[11]See Zimmer (2010) for a discussion of the ethics of Facebook research.

correspond to their offline counterparts. Obtrusive online research methods (surveys, interviews, focus groups, field research and online laboratory experiments) were introduced in this chapter, but they are not the focus of the remainder of this book. In contrast, unobtrusive online research, such as content analysis (both quantitative and qualitative), social media network analysis, online field experiments and online natural experiments, is a major focus of this book. The chapter concluded with an introduction to the ethics of research using digital trace data.

Further reading

Babbie (2007) and Neuman (2006) are classic texts on the practice of social research, and the book by Vogt et al. (2012) aims to assist researchers to choose appropriate methods for different research projects. See also Hewson et al. (2003) and Fielding et al. (2008) for more on online research methods.

3

Social Media Networks

This chapter provides an introduction to social media network analysis, which is the application of network science to the study of behaviour in social media environments.

Section 3.1 introduces to network concepts and terminology. As discussed in Section 1.4, network science is interdisciplinary: Section 3.2 provides an introduction to social network analysis, which is a subfield of sociology. In Section 3.3 it is shown how the examples of online computer-mediated interaction presented in Section 1.2 can be represented and analysed as networks. Threaded conversations, social network services and microblogs are given particular attention in this section as they are the focus of research discussed in Part II of this book.[1] Finally, Section 3.4 discusses how (online) social networks are distinguished from two other types of network: information networks and communication networks.

3.1 SOCIAL NETWORKS: CONCEPTS AND DEFINITIONS

This section presents some key network definitions, which are summarised in Box 3.1 and Box 3.2 (these boxes also include some definitions that are introduced later in the chapter).

A *network* is a set of entities called *nodes* (or vertices) and a set of *ties* (or edges) indicating connections or relations between the nodes. In an *interpersonal network* the nodes are people (often referred to as *actors*) and the edges are connections between people. A *social network* is a type of interpersonal network that is amenable to being studied using an approach called *social network analysis*. Visual representations of social networks are called *sociograms*.

---

[1]Hyperlink network analysis is covered separately in Chapter 4.

BOX 3.1   KEY NETWORK DEFINITIONS

- *Network* – a set of *nodes* and *edges* indicating connections or relations between the nodes.
- *Interpersonal network* – nodes are people (actors) and edges are connections between actors.
- *Social network* – a particular type of interpersonal network, amenable for study using techniques from social network analysis.
- *Dyad* – a pair of actors.
- *Sociogram* – a visual representation of a social network.
- *Graph-theoretic attributes* – node attributes generated from the network, e.g. number of connections or *degree*.
- *Non-graph-theoretic attributes* – node attributes intrinsic to the actor, e.g. sex, race.
- *Directed edge* – has a clear origin and destination and is not necessarily reciprocated.
- *Undirected edge* – represents a mutual relationship with no origin or destination and does not exist unless reciprocated.
- *Unweighted edge* – indicates a binary relationship (a tie either exists or it does not).
- *Weighted edge* – has a value attached to it that indicates the strength of the tie.
- *Isolate* – a node that has no connections to other nodes in the network.
- *Reciprocated edge* – where node i links to or 'nominates' node j, and node j nominates node i.
- *Transitive triad* – a collection of three nodes that are connected to each other in a particular configuration (i nominates j, j nominates k, and i also nominates k).

While some authors may be willing to apply the term 'social network' to any network involving people, this book does not regard all interpersonal networks as social networks. Some interpersonal networks that are formed in social media environments are difficult to be conceived of and studied as social networks. A good example is Twitter: the founders of Twitter emphasise that it is not a social network (in comparison to, say, Facebook) but an information network, and Kwak et al. (2010) confirmed this after studying the patterns of activity within Twitter (for more on this, see Section 3.4).

Network nodes are often represented in sociograms using symbols such as a disc or square, with the choice and size of the symbol reflecting the attribute of the node. There are two major types of node attribute. *Graph-theoretic attributes* are determined from the position of the node in the network (e.g. number of connections or *degree*), while *non-graph-theoretic attributes* are intrinsic to the node (e.g. gender, race, age) and are not related to network position.[2]

Network edges can represent different connections (e.g. collaboration, kinship, friendship), and there are two major types of edges. *Directed edges* have a clear origin and destination (e.g. a Twitter user following another user, a hyperlink from one page to another), and may or may not be reciprocated. A directed edge is represented in a sociogram as a line with an arrow head. *Undirected edges*, in contrast, represent a mutual relationship with no origin or destination (e.g. marriage, Facebook friend), and they do not exist unless they are reciprocated. An undirected edge is represented in a sociogram as a line without an arrow head.

BOX 3.2 TYPES OF SOCIAL NETWORKS

- *Complete network* – contains all the nodes and edges.
- *Personal network* or *egocentric network* ('ego network', 'egonet') – consists of a focal node ('ego') and the nodes connected to the ego ('alters'). There are three subtypes of ego networks:

  — *1.0-degree egonet* only the ties between the ego and the alters are shown (does not show ties between alters).
  — *1.5-degree egonet* – also shows the ties between alters.
  — *2.0-degree egonet* – shows the ties between the alters and also the nodes that the alters are connected to, which are not connected to the ego ('friends of friends').

- *Unimodal network* – a network containing only one type of node.
- *Multimodal network* – a network containing more than one type of node (a *bimodal network* contains exactly two types of node).
- *Multiplex network* – a network with multiple types of edges.
- *Partial network* – created by joining one or more ego networks. A partial network can therefore be used as an approximation to the complete network, with the accuracy of the approximation being dependent on the number of ego networks that are available and how interconnected the network is.
- *Core network* – used in the context of social capital research (Section 7.2.1), and refers to the subset of a personal network containing those people with whom ego discusses 'important matters' and from whom ego can gain social support (generally found via a name generator; see Box 7.3).

---

[2]In mathematics and computer science, the term 'graph' is used synonymously with 'network'; we use the terms interchangeably here.

Finally, edges can also have weights. An *unweighted edge* indicates a binary or dichotomous relationship (i.e. the edge either exists or not). In Facebook a friendship either exists or it does not (pending friendship requests do not count as a tie between the two people), and hence it is represented using an unweighted edge. In contrast, a *weighted edge* has a value attached to it that indicates the strength of the relationship. In a Twitter network a follower edge could be weighted by the number of retweets. In sociograms, weighted edges are generally represented using lines of different width.

Finally, it should be emphasised that the definition of a social network involves three fundamental and interrelated issues (Laumann et al., 1983). First, what constitutes a social tie? Second, who are the actors? Third, where is the network boundary? In the following example of a school friendship network, these three questions are fairly easy to answer, but as we will see in the remainder of this book, they are not always easy to answer in the context of web data.

3.1.1 An example school friendship network

To clarify the above definitions, imagine we are studying the friendships amongst students aged 10–12 years in a small school in a rural area. There are 16 students who all share a single mixed class and there are friendships across different age groups. We interview each student separately and ask them who their friends are. We also collect information on the age and gender of the students. For privacy, we denote each student by a letter (A, B, ..., P).

The sociogram for this network is shown in Figure 3.1 (this is the complete network), where node shape reflects gender (squares are boys, circles are girls) and node size is proportional to the age of the student. One student (P) does not have any outgoing or incoming friendship nominations and is

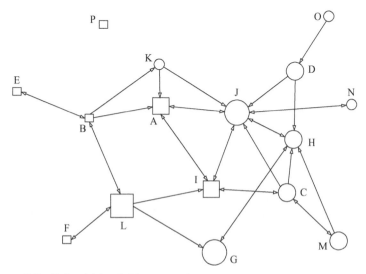

Figure 3.1 School friendship network

known as an *isolate*. We can represent the underlying data using an adjacency matrix (Table 3.1) where a one is recorded in the (i, j) cell if actor i nominates j and a zero otherwise[3]. While adjacency matrices are intuitive methods of representing network data, the fact that adjacency matrices for networks (especially large-scale networks derived from web data) often contain many zeros (they are 'sparse') means that they are an inefficient way of storing network data. An alternative approach is the edge list (Table 3.2).

Table 3.1 Adjacency matrix (partial) for student friendship data

	A	B	C	D	E	F	G	H	I	...
A	0	0	0	0	0	0	0	0	1	...
B	1	0	0	0	1	0	0	0	0	...
C	0	0	0	0	0	0	0	1	1	...
D	0	0	0	0	0	0	0	1	0	...
E	0	1	0	0	0	0	0	0	0	...
F	0	0	0	0	0	0	0	0	0	...
G	0	0	0	0	0	0	0	1	0	...
H	0	0	0	0	0	0	0	0	0	...
I	1	0	1	0	0	0	0	0	0	...
⋮	⋮	⋮	⋮	⋮	⋮	⋮	⋮	⋮	⋮	⋱

Table 3.2 Edge list (partial) for student friendship data

VERTEX 1	VERTEX 2
A	I
A	J
B	A
B	E
B	K
B	L
C	H
C	I
C	J
C	H
D	H
D	J
⋮	⋮

The sociogram displays several features that are common in social networks. First, there is a clustering of boys and girls: this is suggestive of homophily on

---

[3]Throughout this book, lower-case letters are used to denote actors in generic examples, while upper-case letters are used for specific examples (e.g. the school friendship network).

the basis of gender (Section 5.1). Second, a significant proportion of the ties are *reciprocated*, that is, person i nominates person j, and person j nominates person i. This is evidence of the social norm of reciprocity ('returning the hand of friendship'). Third, there are several *transitive triads*, which are triplets of nodes that are connected to one another, e.g. C nominates both J and H as friends, and H nominates J (this is also known as *triadic closure*). Transitivity is another social norm that structures social networks, and this is the idea that 'a friend of my friend is also my friend'. Reciprocity and transitivity reflect social norms that can be derived from balance theory (Heider 1958), which posits that reciprocation of ties and triadic closure can reduce social and psychological strain.

Fourth, some students appear to be occupying significant or important positions within the network. In particular, J receives more friendship nominations than others, that is, there is an unequal distribution of attention (as measured by number of friendship nominations) and this person could possibly exert more *influence* in the social network. This is relevant to several sections in the book, for example Sections 5.2 and 7.1. Further, A and I appear to be positioned between the clusters of girls and boys, indicating that they may be taking a 'bridge-building' or 'boundary-spanning' role in the network, connecting parts of the social network that otherwise would not be connected (Section 9.3.2).

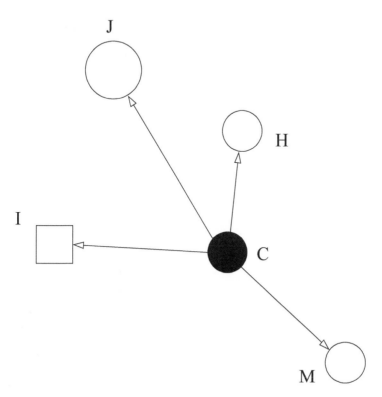

Figure 3.2 1.0-degree ego network (C is the focal node)

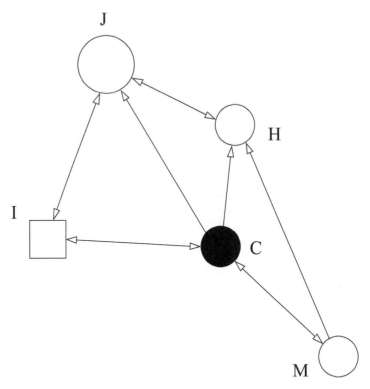

Figure 3.3 1.5-degree ego network (C is the focal node)

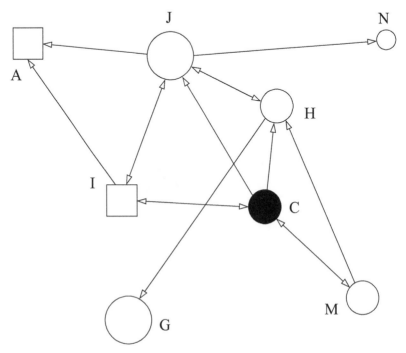

Web Social Science Methods

Figure 3.4 2.0-degree ego network (C is the focal node)

Finally, we can illustrate the different types of networks (Box 3.2) using the school friendship data. Figure 3.2 shows the 1.0-degree egonet for person C, while the 1.5- and 2.0-degree egonets are shown respectively in Figure 3.3 and Figure 3.4.

3.2 SOCIAL NETWORK ANALYSIS

The conceptual framework for defining and analysing social networks comes from social network analysis (SNA), which is a subfield of sociology. We have already been introduced to some SNA terminology above, and this section introduces some of the core concepts and tools in SNA. First, an alternative definition for a social network is provided, one that involves the concept of *social relations*.

3.2.1 Social relations and social networks

A *system* is a set (or collection) of interdependent elements. In biology, a system is a set of species who are interdependent (e.g. predator and prey). Key to a definition of a system is the concept of *boundaries*, which determine what elements are in the system and which are not. A *social system* is a system where the elements are individuals and groups (or actors) in society, and the interdependence between the actors is known as *social structure*.

The concept of social structure has been formalised and developed in the field of SNA. A social network is a network representation of social structure. In a social network, the nodes represent actors in a social system and the ties represent *social relations* between the actors.

The above definition of social network is somewhat different from that in Section 3.1, where we said a social network is a network where the nodes are people and the ties are connections between people. Now we are saying that in a social network the ties are social relations. We define a social relation as a connection between interdependent actors, where this interdependence can work on two levels: tie formation behaviour and actor attributes. Interdependence in tie formation behaviour means that person i's decision to create a tie impacts on j's decision to create a tie. In Section 3.1 we saw two examples of this edge-level interdependency: transitivity ('a friend of a friend is also my friend') and reciprocity ('returning the hand of friendship'). Interdependency of actor attributes means that i's attribute influences j's attribute – for example, j becomes a smoker or changes political viewpoint because of i exerting *social influence* on j (Section 5.2).

The SNA view of the world is that actor interdependence occurs because of the existence of social relations (and in fact a social relation is a tie that reflects interdependence). This immediately reduces the set of types of ties deemed necessary for the existence of a social network. It is clear that friendship can lead to interdependence and hence a friendship network is a social network, but is a citation network a social network? This leads to another potential stumbling block with the analysis of online networks using SNA: when do ties in online networks constitute social relations?

Another way of looking at a social relation is by considering what resources flow through the tie. Two people might be only acquaintances, rather than good friends, but if the tie leads to the flow of information that is useful and is not publicly available, then this can affect behaviour and outcomes and hence should be considered as a social relation. It is useful to note at this point that network nodes do not have to be people for the network to be analysed using SNA. SNA is also used to study, for example, the behaviour of organisations using interlocking board directorates as indicators of ties between firms.

The network perspective

Wellman (1998) proposed five defining features of the 'network perspective':[4]

- SNA differs markedly from other social scientific approaches in that it puts emphasis on the structure of social relations, rather than the attributes of an individual actor, in determining that actor's behaviour and outcomes (see Box 3.3).
- The network perspective focuses on relationships between actors as the unit of analysis rather than the actors themselves. With ordinary least squares – a hallmark technique of 'non-relational' social science – the unit of analysis is the individual. In contrast, with exponential random graph models (Section 3.2.2) – an increasingly used statistical SNA technique – the unit of analysis is the dyad.
- While ordinary least squares involves the assumption that observations are independent (e.g. the error term in a labour market regression model is assumed to be identical and independently drawn from the normal distribution), the network perspective explicitly assumes the *interdependence* of observations (see Section 3.2.2).
- The network perspective recognises that social networks can have both *direct* and *indirect* impacts on individual behaviour and outcomes. Hence, in order to understand the flow of information and resources between two people it is important to know about the entire social network which they are a part of, and not just the fact that they are friends (Box 3.4).
- People tend to belong to several overlapping social networks. That is, individuals tend to be members of several groups and the group boundaries are often fuzzy.

---

[4]See also Katz et al. (2004) for a summary of these features.

BOX 3.3 THE STRENGTH OF WEAK TIES

Imagine Jill and Sue have just graduated from the same university programme and are looking for a job. An economic approach for understanding their labour market success would focus on personal characteristics ('human capital'), such as education, age, labour market experience and skills. In contrast, the SNA perspective is different in two key ways:

- The structure of actor relations both *constrains* and provides *opportunities* to individual actors – social structure can therefore influence the behaviour of actors. Suppose Sue has a wider network of friends and contacts than Jill. Sue is more likely to hear about possible jobs than Jill, with her more limited social network. Thus, Sue's social network is providing opportunities (information) that are affecting her behaviour (applying for a suitable job).
- The structure of social relations can affect outcomes – the likelihood of particular outcomes can be influenced by the structure of the social network. Imagine that Sue and Jill both apply for the same position. Further, imagine that Sue's social network includes people who are known to the hiring committee, and are prepared (possibly informally) to vouch for her; suppose one of the people on the hiring committee asks informally, 'So is this person any good?' Meanwhile Jill does not have any contacts who will vouch for her. We can expect that Sue, who has a good social network, will (all other things considered) get the job.

In the above example, what do we mean by saying that the person with the 'wider network' will have more opportunities to find information on good jobs? We assume that Jill mainly has close friendships with people who also know each other (e.g. they went to the same school, work at the same places), while Sue has more acquaintances in her social network (e.g. people whom she knows, but who do not know each other). Jill's friends all tend to move in the same circles and hence she is unlikely to pick up novel information that can help with her job seeking (e.g. that a firm is looking to hire a person in her area of work). Her friends can thus be thought of as 'redundant ties', in that they are likely to provide the same information. In contrast, Sue's acquaintances, by definition, know people whom she does not know, and Sue is therefore more likely to receive novel information from these ties that can assist with her job seeking.

This idea was formalised by the sociologist Mark Granovetter (1973) in the influential concept of the *strength of weak ties*. Montgomery (1991, 1992) demonstrated that weak ties are positively related to wages and employment rates.

BOX 3.4 STRUCTURAL HOLES IN SOCIAL NETWORKS

The sociologist Ronald Burt has shown that the tendency for people to cluster into social groups on the basis of opportunity structures, such as where they live, their workplace or project team (Section 5.1), leads to gaps or holes in the social structure of communication (*structural holes*), which inhibit the flow of information between people (for a review, see Burt, 1992).

Structural holes can present two types of strategic advantage, relating to *brokerage* and *closure*. Regarding the former, those people who bridge or span structural holes are more likely to gain opportunities for professional advancement because they are exposed to varying opinions, behaviours and sources of information and may be able to combine this disparate knowledge in a way that provides productive advantage. Indeed, it has been shown that such 'bridge spanners' have higher salaries and are promoted more quickly than their peers.

Closure refers to the strategic advantage that is associated with *not* spanning a structural hole, that is, staying within a closed network where there is a high level of connectivity amongst actors. While closed networks do not enable the flow of new information that can provide the 'vision advantage' that is associated with brokerage, they confer another important type of advantage to members: higher levels of trust and coordination (facilitated by high reputation costs for bad or unproductive behaviour, and increased probability that such behaviour can be detected), which can improve team effectiveness and efficiency by lowering labour and monitoring costs.

There are many SNA measures (or *metrics*) that can be used to describe individual actors within a network and the network as a whole – see Section 3.5 for some basic network- and node-level metrics for the example school friendship network.

3.2.2 Statistical analysis of social networks

The above SNA metrics allow us to quantitatively compare social networks and understand the relative position of actors within networks, but a question that often arises is: how did a given social network form?

It is not easy to establish the social processes that led to the emergence of a particular social network. This is because social network data are often collected at a given point in time, and while the data may be collected obtrusively (e.g. by survey or interview), we generally just collect information on whether a tie exists between given actors and not *why* the tie was formed. Instead, we record the social network at a given point in time, and then try to infer something about the preferences of actors based on what we have observed. But this is not an easy task, as there are multiple reasons why a given social tie might be formed.

Network effects

There are two types of features in social networks. *Purely structural* or *endogenous network effects* are network ties that have nothing to do with actor attributes, but arise because of social norms. Section 3.1 introduced two examples of endogenous network effects: *reciprocity* (if i nominates j as a friend, then it is more likely that j will nominate i as a friend), and *transitivity* (if i nominates j, and j nominates k, then there is a higher probability that i will also nominate k, thus forming a transitive triad).

The second major feature in social networks is *actor–relation effects* – these are network ties that are created because of the attributes of actors. There are three main types of actor-relation effects:

- *Sender effects* show the impact of the presence or absence of a particular actor attribute on their propensity to send ties. A significant and positive sender effect indicates that actors with that attribute send more ties than expected by chance.
- *Receiver effects* are analogous to sender effects but refer to the propensity of receiving ties.
- *Homophily effects* show the impact of two actors sharing an attribute on the likelihood of there being a tie between the two actors.

We can illustrate these ideas by looking at Figure 3.5, which shows a transitive triad that has been extracted from the example friendship

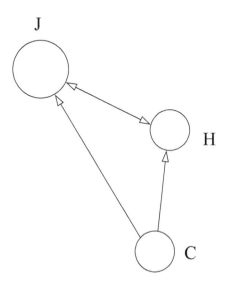

Figure 3.5 Transitive triad extracted from school friendship network

network depicted in Figure 3.1. Why has this transitive triad formed? In particular, why has C nominated J as a friend? There are three potential reasons:

- C may have nominated J as a friend because they are both girls (this would be an example of homophily).
- C may have nominated J as a friend because J is older and C would like to be seen to be 'hanging out with the older girls' (this would be an example of a receiver effect, with J receiving more nominations because she is older).
- The tie from C to J could also be a purely structural effect: the fact that C nominates H and H nominates J means there is a higher likelihood that C will also nominate J (thus forming a transitive triad).

The main point here is that it is very difficult to distinguish actor–relation effects from purely structural effects when the social network is observed at a single point in time.

Exponential random graph models

We now briefly introduce an important and relatively new statistical SNA approach, exponential random graph models (ERGMs) or $p\star$ models, which are a particular class of statistical model for social networks (Frank and Strauss, 1986; Wasserman and Pattison, 1996; Pattison and Wasserman, 1999; Robins et al., 1999). While a detailed introduction to ERGMs is beyond the scope of this text, it is useful to understand what they are designed to do.[5]

ERGMs are a statistical technique that enables the explicit modelling of the dependence among the units of observation, which are dyads. Suppose we have three people: Ann, Sue, David. Assume that Ann and Sue are friends and Ann and David are friends. Earlier statistical approaches to modelling social networks involved the implausible assumption that the probability of Sue and David forming a friendship is the same in this situation as it would be if Ann was not friends with either of them (thus ignoring a basic mechanism in social behaviour, triadic closure).

An ERGM is essentially a pattern recognition device which breaks a given network down into all of its constituent network *motifs* or *configurations* and then tests whether particular configurations occur more (or less) frequently than would be expected by chance alone.[6] In a manner similar to standard regression techniques, the model estimation produces

---

[5]See, for example, Robins et al. (2007) for a detailed introduction.

[6]Faust and Skvoretz (2002; and Skvoretz and Faust (2002)) referred to these as *structural signatures*.

parameter estimates and associated standard errors. If a particular network motif occurs at greater or less than chance levels, we can then infer that the associated social relation has had a significant role in the development of the social network.

ERGMs therefore allow the researcher to statistically identify various purely structural and actor–relation network effects. Not controlling for purely structural self-organising network properties may lead to a spurious conclusion that the attributes of actors are driving social tie formation when in fact it is purely structural self-organisation. In the context of homophily research, for example, ERGMs thus enable researchers, in effect, to 'control' for purely structural effects such as reciprocity and transitivity, and thus accurately measure the homophily actor–relation effect.

There are two main classes of ERGMs: social selection models and social influence models. With social selection models, the behaviour that is being explained is tie formation: tie formation is therefore modelled as depending on the distribution of node attributes and the overall structure of the network. An ERGM of homophily in a friendship network is an example of a social selection model. With social influence models, the aim is to explain one of the attributes of nodes, using other attributes and the structure of the network as explanatory variables. An example of a social influence model would be an ERGM of smoking behaviour of teenagers, looking at the role of individual and household-level characteristics (e.g. income or education of parents) and the structure of social relations (peer effects).

3.3 SOCIAL MEDIA NETWORKS

In Section 1.2 seven types of online interaction were introduced: threaded conversations, Web 1.0 websites, blogs, wikis, social network sites, microblogs and virtual worlds. This section first shows how each of these forms of online interaction can be represented as social networks and then provides more detail on threaded conversations, social network sites and microblogs.

3.3.1 Representing online interactions as interpersonal networks

Threaded conversations

In a newsgroup, person A starts a new thread by submitting a post 1 at time t. Person B responds with post 2 at time $t + 1$. At time $t + 2$ person C submits post 3 to the thread, which is a response to B's response. At time $t + 3$ person

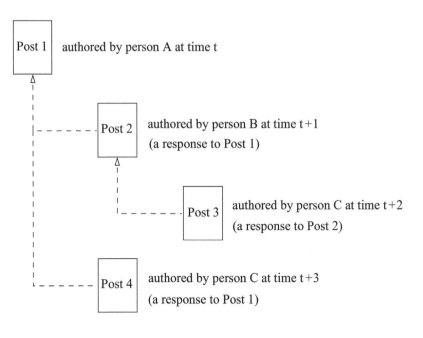

authored by person A at time t

authored by person B at time t+1
(a response to Post 1)

authored by person C at time t+2
(a response to Post 2)

authored by person C at time t+3
(a response to Post 1)

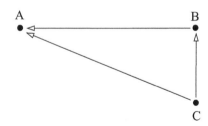

Figure 3.6 Threaded conversation network

C also makes a direct response to the initial post. This activity can be represented as a directed and unweighted network between the three people (Figure 3.6).

Web 1.0 websites

Two organisations (A and B) each have a website. The webmaster for organisation A creates a hyperlink from a webpage on his organisation's website to a webpage on the website of organisation B. This can be represented as a directed and unweighted network between the two organisations (Figure 3.7).

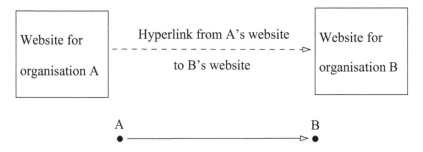

Figure 3.7 Web 1.0 network

Blogs

Two people (A and B) each blog on a similar topic. Blogger A writes a blog post where he comments on a post that was earlier written by blogger B, and to assist the reader he creates a hyperlink pointing to the earlier blog post. This can be represented as a directed and unweighted network between the two bloggers (the diagram describing this is analogous to Figure 3.7). Note that to capture the temporal nature of blogging, we could have also represented this behaviour using Figure 3.6.

Wikis

Person A and person B both edit wiki page 1. Person B also edits wiki page 2, which is also edited by person C (Figure 3.8). At this point, it is useful to introduce some additional social network terminology. A *unimodal network* contains only one type of node, while a *multimodal network* contains more than one type of node. The wiki network just described is an example of a bimodal network containing exactly two types of vertices: wiki editors (people) and the wiki pages they edit (this is also known as an affiliation network). So, in this particular affiliation network, people do not connect directly with people and wiki pages do not connect directly with wiki pages. However, a bimodal affiliation network can be transformed into two separate undirected and unweighted unimodal networks (wiki editor to wiki editor and article to article), as is shown in Figure 3.8.

Social network sites

Two people (A and B) have profiles on Facebook. Person A requests person B to become a Facebook friend, and person B accepts the friendship request. This can be represented as an undirected and unweighted network between the two people (Figure 3.9).

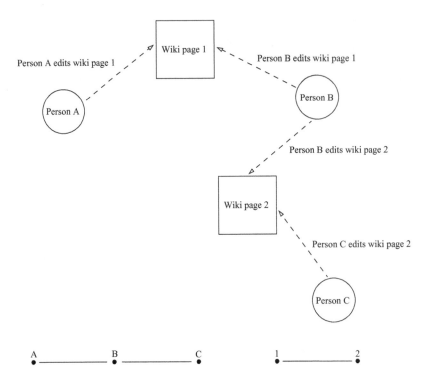

Figure 3.8 Wiki network

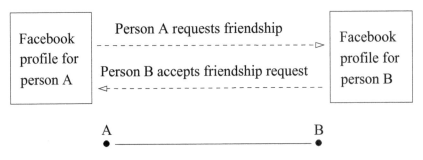

Figure 3.9 Facebook network

Microblogs

Person A follows person B on Twitter (see Section 3.3.4 for more on Twitter terminology). Person C authored a tweet where person B was mentioned. Therefore, there are two types of directed edges (follows, mentions) and this is an example of a *multiplex network*, which is a network with multiple types of edges. As noted in Section 3.3.4, with Twitter there are actually three types of directed edges: following relationships, 'reply to' relationships and 'mention' relationships. Multiplex ties are often reduced to a simplex tie (e.g. a tie exists if any of the multiplex ties exist). Thus we can represent this example multiplex Twitter network as a directed and unweighted simplex network with person A nominating person B, and person C nominating person B (Figure 3.10).

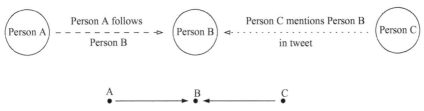

Figure 3.10   Twitter network

Virtual worlds

Three people are playing World of Warcraft. Person A and person B both join raiding party 1 (parties allow players to work together as a team, enabling private in-game communication and sharing of resources). Person B also joins raiding party 2, which person C also belongs to. This can be represented as an affiliation network with exactly the same structure as the wiki network in Figure 3.8 (and the two unimodal networks for people and raiding parties).

3.3.2 Threaded conversations

Online newsgroups and forums enable a form of *threaded conversation* which can be analysed as a network, and the present section provides an introduction to so-called *thread networks*.[7]

A threaded conversation can be represented as a directed network with a directed tie representing someone replying to another person's message. The following key properties of threaded conversations have been identified (e.g. Resnick et al., 2005):

- A set of *topics* or groups where the threaded conversation occurs.
- Within each topic there are *threads*: top-level *posts* (or messages) and responses to those posts (and, generally, responses to responses).
- Each post in a thread is authored by a single person.

---

[7]This section draws from Hansen et al. (2010b). See also Vergeer and Hermans (2008) for more on using content analysis and network techniques to analyse online political discussions.

- Posts are typically permanent – they cannot be edited after posting.
- Users are generally presented with the same view of the messages in a topic (e.g. chronological, reverse chronological).

Figure 3.6 showed a basic example of a thread network. Here, we use a different approach to displaying threaded conversations in order to further illustrate thread networks.

There are two main types of threaded conversation topics. A *discussion topic* is where people creating threads pose questions or statements that are meant to generate discussion, with contributors or posters expressing various perspectives and opinions. An example of a discussion topic is the comp. os.linux.advocacy newsgroup, which is devoted to discussing the benefits of Linux compared to other operating systems.[8] The following is a sample of recent thread titles in this newsgroup, as at 8 January 2012:

- What sort of person 'admires' companies?
- 'Advocates' ... listen up, Good News: German cities following Munich's open source example
- Replacing Windows with GNU/Linux

The threads in discussion topics are typically quite long: the 20 most recently updated threads in the comp.os.linux.advocacy newsgroup (as at 8 January 2012) had an average of 34 posts.

The second type of threaded conversation topic is called a *question-and-answer topic*, and it typically involves threads with questions that elicit straightforward answers. Newsgroups or forums that are focused on particular software are good examples of question–and–answer topics. For example the support forum for the open source Mozilla Thunderbird email software[9] has the following recent thread titles (as at January 2012):

- ver. 9.0.1 will not open a mail window from another program
- Thunderbird 8 and 9 chooses wrong printer
- What is the Profile folder?

The threads in question–and–answer topics tend to be short, since they typically involve someone asking a question and others providing an answer. The 20 most recently updated threads in the Mozilla Thunderbird support forum (as at 8 January 2012) had an average of only 5 posts.

Figure 3.11 shows posts from an example discussion topic. The posts are organised chronologically, with the level of indent reflecting the position of the post in the thread. Person A started Thread 1 in the example discussion topic, and B and C both responded to this initial post. D then responded to

---

[8]http://groups.google.com/group/comp.os.linux.advocacy/topics

[9]http://forums.mozillazine.org/viewforum.php?f=39

Thread 1 | Person A | 9.00am 08/1/2012

Re. Thread 1 | Person B | 9.10am 08/1/2012

Re. Thread 1 | Person C | 11.30am 08/1/2012

Re. Re. Thread 1 | Person D | 2.00pm 09/1/2012

Re. Re. Re. Thread 1 | Person B | 4.00pm 09/1/2012

Thread 2 | Person E | 1.30pm 10/1/2012

Thread 3 | Person D | 10.00am 11/1/2012

Re. Thread 3 | Person C | 10.00pm 11/1/2012

Figure 3.11 Posts in a discussion topic

C's post, and, in turn, A contributed another post to the thread (a response to C's post). Two other threads were also started in this topic.

An example question-and-answer topic is shown in Figure 3.12. All three threads consist of an initial post and then one person following up with a response, and that is the end of the thread.

There are two main ways a threaded conversation can be represented as a network. In a *reply network*, when person i replies to person j's post, a directed tie is created from i to j. Multiple replies increase the weight of the tie. In a *top-level reply network*, all repliers are connected to the person who started the thread (and not the person they were replying to, unless that

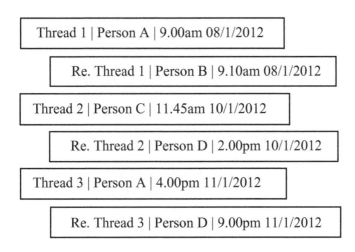

Thread 1 | Person A | 9.00am 08/1/2012

Re. Thread 1 | Person B | 9.10am 08/1/2012

Thread 2 | Person C | 11.45am 10/1/2012

Re. Thread 2 | Person D | 2.00pm 10/1/2012

Thread 3 | Person A | 4.00pm 11/1/2012

Re. Thread 3 | Person D | 9.00pm 11/1/2012

Figure 3.12 Posts in a question-and-answer topic

person started the thread), and hence, thread starters are more prominent. In online newsgroups and forums where threads are typically short (e.g. technical topics where people ask questions and get answers), the top-level reply network better reflects the dynamics of the question-and-answer topic. However, in online newsgroups and forums where a lot of discussion occurs, the reply network will be a better representation since people are often posting replies to replies. The posts from the example discussion topic in Figure 3.11 are shown as both a reply network and top-level reply network in Figure 3.13.

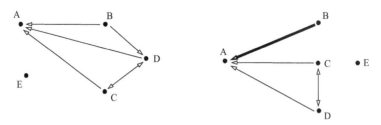

Figure 3.13 Discussion topic threads represented as reply network (left), top-level reply network (right)

Data sources and tools

Archives of Usenet discussion groups are freely available. For example, in 2001 Google started Google Groups and offered 20 years of Usenet data dating back to 1981 (more than 800 million messages).[10] Archives such as Google Groups are searchable via a web browser; however (as with the Wayback Machine of the Internet Archive, discussed in Section 4.3.2), they are not currently suitable for large-scale empirical analysis.

In order to do large-scale empirical research of a Usenet group, it is necessary to get access to the underlying repositories of emails. Typically a repository for a Usenet group consists of a set of text files, where each text file contains all of the emails sent to the group in a particular period (e.g. month). Text parsing routines can then be used to parse the emails, using email header information and thread markers (it is considered 'netiquette' when responding to a previous post to include a 'snippet' of the post in your email).

Generally, if you are going to work with Usenet data, it is best to work with a dataset that has already been extracted from the raw data. Two examples of projects that focused on extracting datasets from Usenet are Netscan and SIOC (Box 3.5).

Web Social Science Methods

---

[10]See http://www.google.com/googlegroups/archive_announce_20.html

3.3.3 Social network sites

Social network sites such as Facebook, Myspace and LinkedIn are increasingly being used for social science research. Social network sites have opened up new research possibilities for social scientists, and their introduction has overcome some of the challenges associated with using web data (in particular, hyperlink network data) for research into connected social behaviour.

First, from a conceptual point of view, with social network sites the nodes are clearly people and the network ties are also relatively easily interpreted. While it is obvious that Facebook friendships are qualitatively different from offline friendships (in terms of the cost of making and maintaining the tie, for example, and in terms of the public–private nature of the action), it is easier to interpret a tie in Facebook than a hyperlink tie between static websites.

Second, the relatively easy interpretation of network nodes and ties also means that social science research into online social networks is less methodologically challenging, compared with hyperlink networks. Users of social network sites are encouraged to describe themselves (e.g. political persuasion, religion, humour, smoking and drinking behaviour) using a combination of text fields, drop-down selection boxes and check boxes. The profiles are highly amenable to automated data analysis (compared with web pages which are much less structured).

The successful social network sites are commercial enterprises and the owners realise the marketing value of user data. For this reason, it is not easy to get access to large-scale datasets from social network sites (crawling these sites is generally not permitted). However, some researchers have managed to get access to large-scale datasets from Facebook (Section 5.1.2). The use of Facebook applications also enables researchers to collect data on particular subsets of users (see below, and also Section 10.2.1).

Lewis et al. (2008) accessed the profiles of all incoming students at a US college, and tracked these students throughout the four years of their college education.[11] The authors used both Facebook profiles and college records to collect socio-economic data (gender, race, ethnicity, median household income, region of origin), and cultural preference data (preferences for music, books and movies) were also collected from the Facebook profiles. Three measures of social tie were collected: being Facebook friends, being Facebook picture friends (which the authors regard as indicating a 'higher level of positive affect towards alter compared to a Facebook friend' (p. 333)), and being roommates.

Lewis et al. (2008) contend that their Facebook dataset has five defining features:

- *Natural research instrument.* The advantages are that interviewer effects, recall error and other sources of measurement error typically associated with surveys are minimised. Further, there is no need (from a cost perspective) to restrict the size of the reported social network. However, the authors had to impose a boundary on the network since they did not have access to the Facebook profiles of people outside the college who were Facebook friends with the college students. Lewis et al. (2008) argue that this boundary did not impact on their research since their focus is on relationships 'most relevant for the conduct of everyday life at this (residential) campus' (p. 331). The disadvantages of their natural research instrument are uncertainty over interpreting the attribute and relational data (what is the exact meaning of a Facebook friend?); and that the cultural data reflect both true preferences and strategic presentation of self.
- *Complete network data.* The authors were able to collect complete network data thus allowing the construction of node-level metrics (e.g. degree, betweenness and closeness centrality) and network-level metrics (e.g. density, centralisation). The disadvantages are that complete networks are not representative of the wider population, but this is not particular to research using Facebook data – this is a potential limitation of SNA in offline settings too.
- *Longitudinal data.* The authors have four waves of data following the students through college. The advantage of this is that it will shed light on how networks and tastes co-evolve. A disadvantage is that people tend to create Facebook friendships but not terminate them (and hence Facebook is better for studying tie formation than tie dissolution). Another issue is

---

[11]There are other examples of research into friendship formation that use data from social network sites (see, for example, Mayer and Puller, 2008; Thelwall, 2009a).

that 220 users changed their profiles from public to private, and this is problematic if attrition is related to the behaviour being studied.

- Multiple social relationship. Three measures of relationships (mentioned above).
- Cultural preference data. Via Facebook the authors were able to collect extensive data on tastes towards, e.g. music, movies and books.

Boyd and Ellison (2008) argue that social network sites are different from earlier forms of online communities such as newsgroups because they primarily support pre-existing social relations (e.g. maintaining or solidifying existing offline relations rather than being used to meet new people). Lampe et al. (2006) found that Facebook users tend to search for people with whom they already have an offline connection rather than browsing for complete strangers to meet. Lenhart and Madden (2007) found that 91% of teenagers in the USA using social network sites do so to connect with offline friends. Research has therefore shown that people generally use Facebook to connect with friends they have in the 'real world', hence Facebook data are useful indicators of real-world friendships. As Wimmer and Lewis (2010, p. 585) note with regard to Facebook picture friends: 'Online pictures document an existing "real-life" tie and are therefore qualitatively different from the "virtual" network studied by other researchers.'

Data sources and tools

As noted above, obtaining large-scale datasets from commercial social network sites such as Facebook is difficult and researchers are left with two options (see Ackland, 2009). First, it might be possible to set up a niche social network site targeted at a particular community of people; this is what Centola (2010) did in his study of social networks and health behaviour (see Section 5.2.2). The advantage of this is that since the researcher has control over the social network site, then the data are available. The obvious disadvantage is that you need to convince people to use the site.

The second option, which is possibly more viable given the dominance of commercial social network sites (in particular, Facebook), is to develop a Facebook application that allows Facebook users to opt into a research project (i.e. make their profiles and social networking activities visible to researchers). This is still an example of unobtrusive web research, because while the Facebook user has given consent to the research (by installing the Facebook application), they do not interact with the researchers; they just keep on using Facebook as they would normally, but the researcher is able to collect data on their activities (e.g. updates to profiles, friending behaviour). Aral and Walker (2010) took this second approach in their study of social influence in Facebook (Section 10.2.1). Also relevant here is NameGenWeb, which is a tool for constructing ego networks using data from Facebook (Hogan, 2010).[12]

---

[12]http://namegen.oii.ox.ac.uk/namegenweb/learnmore.php

3.3.4 Microblogs

The underlying technology for microblogs is quite similar to blogging (Section 4.3.3), but posts ('tweets', as they are called by Twitter) are limited to 140 characters. While there are several examples of microblogs, Twitter is the best known and so the following discussion is focused on Twitter (many of the features described here are common to other microblogs).

Tweets are typically public and 'one-to-many'. So how can we extract or identify directed ties between particular Twitter users? There are four types of interactions between Twitter users that can be represented as network ties.

First, it is possible to subscribe to or *follow* another Twitter user and you will then receive that person's tweets. The people you are following are your *friends*, while the people who follow you are your *followers*.

It is also possible to directly include another Twitter user in the body of a tweet. This may be a better indication of the existence of a social tie between two users than friends or follower lists. In the same way, it is argued by some that in blogs, permalinks (hyperlinks in blog posts) are a better indication of connections between bloggers than blogrolls, which can be notoriously stale (Section 4.3.3).

@replies are when the tweet starts with a username (prefixed with '@'), and are used to show that a tweet is intended for a particular user. @replies can be contrasted with *@mentions*, which is where the username (prefixed with '@') appears somewhere in the tweet, but not at the start. So the tweet 'We saw @rob at the football' is an @mention, while '@rob – was that you we saw at the football?' is an @reply. So @mentions indicate acknowledgement while @replies show that a tweet is being directed to a particular user.

The fourth type of interaction in Twitter is the *retweet* function, which is where a Twitter user forwards a tweet to their followers, prefixing the tweet with 'RT @user', where 'user' is the person who authored the original tweet. So, if Twitter user @dave tweeted 'We saw @rob at the football' and one of his followers decided to retweet this, this would be done using 'RT @dave: We saw @rob at the football'. Retweeting is a sign that the person doing the retweet thinks the original tweet is interesting or important enough that it needs to be read by a wider audience (obviously not so in this sample), and hence is a sign of acknowledgement.

As discussed below, there are tools that can be used to find, for a given user, who they follow and who they have @mentioned, @replied to or retweeted, thus allowing the construction of a directed multiplex 1.0-degree egonet. By repeating this process for each of the users or alters identified in this way, it is possible to create 1.5- and 2.0-degree egonets. By repeating this process for egos, it is possible to create a complete network showing the follower relationships amongst the Twitter users. In some contexts it may make sense to just focus on one type of tie (e.g. follower) or else to convert the multiplex network into a simplex network.

In terms of identifying the boundaries to a given Twitter network, the Twitter Application Programming Interface (API) allows you to identify all the users (over a particular period of time) who have authored tweets on a particular topic. This can be done via the Twitter *hashtag* (#) which is the convention for identifying the topic of a tweet, or else by searching for a particular word or phrase. So it would be possible to construct the network of Twitter users who tweeted on '#climatechange' or 'climate change', for example.

Data sources and tools

There has been a lot of research interest in Twitter, largely fuelled by the fact that Twitter have had a relatively open policy towards their data. The Twitter Search API[13] presently allows the extraction of historical tweets containing particular hashtags or text strings (how far back the search goes is variable, but is approximately up to a week). The Twitter API can also be used to find the friends of a given Twitter user.[14] As discussed further in Section 4.3.1, accessing data via an API allows software (termed 'clients') to can programmatically retrieve data from remote servers. An example of client software for using the Twitter API is the Python tool tweepy,[15] and the social media analysis tool NodeXL[16] also has a plug-in for accessing data via the Twitter API. Researchers (e.g. Kwak et al., 2010; Yang and Leskovec, 2011) have also constructed large-scale Twitter datasets which, until recent changes in the Twitter Terms of Service, were freely available to the research community.

3.4 SOCIAL NETWORKS, INFORMATION NETWORKS AND COMMUNICATION NETWORKS

It is useful to think of a hierarchy of interpersonal networks that can be studied in social media environments:[17]

- An *affiliation network* is where ties between people do not indicate a direct flow of information. For example, it is possible to construct an affiliation network of people who have edited the same page in Wikipedia, but they will not necessarily know each other or pass information between themselves.

---

[13]https://dev.twitter.com

[14]At the time of writing, Twitter are planning to phase out the Search API in favour of the Streaming API, which allows researchers to compile their own databases of historical Twitter data by accessing near real-time flows of tweets (this is referred to as accessing the Twitter 'firehose').

[15]http://tweepy.github.com

[16]http://nodexl.codeplex.com

[17]See also Park (2003) who compares social networks, information networks and hyperlink networks.

- An *information network* involves one-way transmissions of information between actors, where the information being sent by one actor is being broadcast to many actors. A good example of an information network is Twitter. Kwak et al. (2010) studied the Twitter follower graph and concluded that on the basis of the highly skewed distribution of followers (see more on power laws in Section 7.1.1) and the low reciprocation rate, Twitter more closely resembled an information network than a social network.
- A *communication network* involves a two-way exchange of information between actors, where the information being sent by one actor is intended for another particular actor.
- A *social network* involves two-way exchange of information (or at least the *potential* for two-way exchange of information) and interdependence between the actors, in that actor i's behaviour (tie formation or other behaviour) can potentially affect j's behaviour. That is, there is potential for i to exert social influence over j.

3.4.1 Flows of information and attention

We have discussed different ties in social media networks but have not mentioned what flows along these ties. Is it important to do so? With an offline friendship network, researchers will not necessarily specify what is flowing along the ties, but it can be inferred by the nature of the research. For example, to what extent does social influence contribute to teenager smoking (e.g. Mercken et al., 2010)? Is obesity 'socially contagious' (e.g. Christakis and Fowler, 2007)? The implication is that information is flowing across the tie, leading to an actor changing behaviour.

In the context of social media, we need to pay much greater attention to the question of what flows across ties. This is because it helps us to classify what the network is and how it should be studied. Ties in social media networks can be classified as either facilitating flows of attention or flows of information (and in some cases both). This can be clarified with some examples.

First, consider a dyad from a Facebook friendship network (person i and person j are Facebook friends). There is a flow of attention – i is publicly declaring that j is a friend, and vice versa. There is also likely to be a flow of information (they get access to each other's Timelines and status updates, for example), but researchers typically will not be able to know if i has received information from j via a status update, for example, or via a private message. So it would be difficult for researchers to know whether this is an information network. Is this a social network? Yes, if only because many Facebook friends are real-world friends (Section 3.3.3). This would lead us to conclude that it is reasonable to study this network using SNA. Is this an information network? Not that we can observe. Is this a communication network? Not that we can observe.

The second example is that of a dyad from a thread network (person i responds to person j's forum post). There is a flow of attention from i to j: i is publicly acknowledging having read j's post and is paying enough attention to j to want to respond. Is there a flow of information? Yes, we can assume that in order to post a reply to j, i must have read the original post and hence consumed that information. This is definitely an information network then. But is it a communication network? No – it is information transmission, rather than exchange (one-to-many, not one-to-one). Since it is not a communication network, then it would be hard to categorise this as a social network.

Blog networks and Twitter networks can be thought of in the same way as thread networks. That is, as they are typically used, they are most likely to be information networks, but not communication or social networks.

3.5 SNA METRICS FOR THE EXAMPLE SCHOOL FRIENDSHIP NETWORK (ADVANCED)

In this section, basic node- and network-level metrics are calculated for the subnetwork presented in Figure 3.3.

3.5.1 Node-level SNA metrics

- *Node indegree* is the number of inbound ties for a given node (note: only defined for directed networks). For our example network, the indegree for node C is Z, while the indegree for node H is 3.
- *Node outdegree* is the number of outbound ties for a given node (note: only defined for directed networks). For our example network, the outdegree for node C is 4, while the outdegree for node H is 1.
- *Node degree* is only defined for undirected networks and is equal to the number of ties that the node has.
- *Betweenness centrality* gives an indication of the extent to which an individual node plays a 'brokering' or 'bridging' role in a network and is calculated for a given node by summing up the proportion of all minimum paths within the network that 'pass through' the node. Betweenness is sometimes normalised to sum to 1 over all nodes.

  For our example network, the betweenness centrality score for node C is calculated in the following way. First, find all ordered-pair of actors (dyads) that do not include C; these dyads are (H, I), (H, J), (H, M) and (I, H), (I, J), (I, M), (J, H), (J, I), (J, M), (M, H), (M, I) and (M, J). There is one shortest path between H and I, and this does not pass through C (H and I are directly connected), so the proportion of shortest paths between H and I that pass through C is 0. So, for the 12 dyads, the proportions of minimum paths that pass through C are: 0, 0, 1, 0.5, 0, 1, 0, 0, 1, 0 , 1, 0.5. Summing up these 12 proportions gives 5 – this is the betweenness centrality score for C. In contrast, the betweenness centrality score for I is 4, indicating that C is more central in this network (but recall that this is the 1.5-degree ego network for C, so this result is expected).

- *Closeness centrality* is an indicator of the extent to which a given node has short paths to all other nodes in the graph and it is thus a reasonable measure of the extent to which the node is in the 'middle' of a given network. Closeness is sometimes normalised to sum to 1 over all nodes. Typically, closeness centrality is only calculated for networks that are *strongly connected*, meaning that every node can reach all other nodes (either directly or indirectly).[18] The reason for this is that in a network that is not strongly connected there will be some nodes that have an infinite length of the minimum path between them (since they do not connect to one another either directly or indirectly). Our example network is strongly connected, so closeness centrality can be computed. The closeness centrality for C is $(5-1) \div (1+1+1+2) = 1$, where the numerator is network size minus one, and the denominator is the sum of the geodesic distances 'number of jumps' between the dyads (C, H), (C, I), (C, J) and (C, M), respectively. The closeness centrality for I is lower at 0.67, again indicating that C has a more central position within this network.

3.5.2 Network-level SNA metrics

- *Network size* is the number of nodes in the network. For our example network, the network size is 5.
- *Network density* is the number of network ties as a proportion of the maximum possible number of network ties. The maximum number of ties in a directed network of size n is $n(n-1)$, while for an undirected network the maximum number of ties is $n(n-1)/2$. So for our example network (which is directed), the maximum number of ties is 20, and the network density is $11/20 = 0.55$.
- *Network inclusiveness* is the number of non-isolates as a proportion of the network size. For our example network there are no isolates and so network inclusiveness is 1.
- *Centralisation.* Centralisation is a network-level property and it broadly measures the distribution of importance, power or prominence amongst actors in a given network. In effect, centralisation measures the extent to which the network revolves around a single node or small number of nodes. The classic example of a highly centralised network is the star network where the node in the centre of the star has complete centrality while the other nodes have minimal centrality. In contrast, a circle network is highly decentralised since all nodes share the same centrality.

---

[18]A *connected component* is a set of nodes that are connected (either directly or indirectly) to one another. A *strongly connected component* is where each node is reachable by all other nodes.

Centralisation is calculated by first calculating a particular node centrality measure and then finding the sum of the absolute deviations from the graph-wide maximum (generally, the centralisation score is also normalised by the theoretical maximum centralisation score).

For our example network, we can calculate the following unnormalised centralisation measures (ordering the nodes by C, H, I, J and M):

- Unnormalised *indegree centralisation* is
  $(3 - 2) + (3 - 3) + (3 - 2) + (3 - 3) + (3 - 1) = 4$.
- Unnormalised *outdegree centralisation* is
  $(4 - 4) + (4 - 1) + (4 - 2) + (4 - 2) + (4 - 2) = 9$. The network is therefore more centralised on the basis of outdegree, compared with indegree.
- Unnormalised *betweenness centralisation* is
  $(5 - 5) + (5 - 0.5) + (5 - 4) + (5 - 3.5) + (5 - 0) = 12$.
- Unnormalised *closeness centralisation* is
  $(1 - 1) + (1 - 0.4) + (1 - 0.67) + (1 - 0.57) + (1 - 0.67) = 1.7$.

3.6 CONCLUSION

This chapter has provided an introduction to social network analysis and has shown how SNA concepts and tools can be used (and adapted) for studying social media network data. Particular focus was given to three types of social media data: threaded conversations, social network services and microblogs (hyperlink networks are covered separately in Chapter 4). Many of the concepts in this chapter are used in the remainder of the book.

Further reading

The canonical social network analysis reference is Wasserman and Faust (2004). An introduction to SNA that is available online is Hanneman and Riddle (2005). For more on analysis of communication networks, see Monge and Contractor (2003), and for a treatment from the network science (applied physics/computer science) and economics perspectives, see Easley and Kleinberg (2010). For an introduction to social media network analysis, see Hansen et al. (2010a).

4

Hyperlink Networks

This chapter provides an introduction to hyperlink network research, with the main focus being on Web 1.0 websites – the Static Web (Box 1.3).[1]

When we think of Web 1.0, we are typically focusing on organisations (e.g. corporations, government agencies, non-government organisations) that are maintaining websites. Section 4.1 contains a discussion about why organisations create hyperlinks and the benefits of receiving hyperlinks. Section 4.1 also introduces three fundamental questions about defining a hyperlink network (what the tie is, what the nodes are and where the boundaries are), drawing on the related discussion in the context of social networks from Section 3.1. Section 4.2 outlines three disciplinary approaches to social scientific research into hyperlink networks: hyperlink networks as citation networks (information sciences), hyperlink networks as issue networks (media studies), and hyperlink networks as social networks (sociology). In Section 4.3 there is an introduction to tools for hyperlink retrieval (web crawlers) that have been used in humanities and social science research.

4.1 HYPERLINK NETWORKS: BACKGROUND

The web is a vast electronic library of hyperlinked documents, but the social scientific approach to studying hyperlinks involves conceptualising the web as being something other than simply a repository of electronic information.[2] As Jackson (1997) noted: 'Once we become critical of the assumption that the Web is a neutral repository of information, the structure of the Web becomes much more interesting.'

However, in order to study the structure of the web from a social scientific perspective, it is important to think about why organisations create hyperlinks (and the benefits they gain from receiving hyperlinks) and to consider

---

[1] The technology that underlies Web 1.0 websites is very similar to that of blogs, and there is a discussion about the latter in Section 4.3.3.

[2] In this chapter, references to the web are to be taken as meaning 'Web 1.0'.

the key methodological question of how to construct a hyperlink network for use in research.

4.1.1 Motives for sending, and benefits of receiving, hyperlinks

Social science web research often involves the study of groups, organisations or companies where we are trying to learn something about what it means to receive or send a hyperlink from the perspective of organisational behaviour. In the context of unobtrusive research, where we are using a web crawler to collect hyperlink data without either asking the website owners why they created the link or looking at the text surrounding the hyperlink, it is not straightforward to know what is being exchanged by a hyperlink.

Sending hyperlinks

The hyperlink is commonly seen as the 'essence' of the web (Jackson, 1997; Foot et al., 2003). Several possible interpretations have been offered for why a hyperlink might be created or sent. Hyperlinks can be seen as 'conferrers of authority' or endorsement (Kleinberg, 1999), indicators of trust (Davenport and Cronin, 2000), reflections of organisational communicative and strategic choices (Rogers and Marres, 2000), and as tools of organisational alliance building and message amplification (Park et al., 2004).

The existence of so many interpretations of the meaning of a hyperlink between two websites has led Thelwall (2006) to conclude that there can be no single 'theory of linking'. Certainly, the motivation for hyperlinking will vary depending on the context – a government department, for example, will have a different motivation for hyperlinking than a political party.

In Section 6.2.1 we discuss Shumate and Dewitt (2008), who conceptualise the hyperlinking activities of non-government organisations (NGOs) focused on HIV/AIDS as being directed towards the creation of an information public good (a hyperlink network that enables information on this issue to be located). In contrast, Ackland and O'Neil (2011) regard the hyperlinking behaviour of activist organisations as being related to the formation of online collective identity (Section 6.3). In Section 8.1 the number of outbound links from a government department website is seen as a measure of 'extroversion', providing a quantitative measure of the prominence or centrality of the website in social and information networks, also known as 'nodality' (Hood, 1983; Hood and Margetts, 2007).

Receiving hyperlinks

With regard to the receipt of hyperlinks, as discussed further in Section 7.1.1, inbound links are important in driving traffic to websites for two reasons. First, the more inbound hyperlinks from other relevant websites, the greater the number of pathways that people can follow to the website. Second, inbound hyperlinks are a primary determinant of a site's ranking on search engines such as Google. As put by Hindman et al. (2003), website *retrievability* is an absolute concept (if the website is 'down' the content is not viewable, but as long as it is 'up' it is as viewable as the content on any other website). Website *visibility*, however, is relative, and is largely determined by the number of inbound hyperlinks from other relevant websites.

In Section 7.1, the importance of receiving hyperlinks is discussed in the context of the visibility of political information. Section 9.2.1 considers whether counts of inbound hyperlinks can be used as scientometric measures of academic authority and output.

4.1.2 Hyperlink network nodes, ties and boundaries

As noted in Section 3.1, the fundamental methodological challenge in social network research is defining the nodes, ties and boundaries to the network. We now look at each of these in the context of hyperlink networks.

Hyperlink network nodes

Defining the nodes in a hyperlink network can be more complicated than with other types of online networks. The nodes in Facebook and Twitter networks are easy to define, since each username in these social media environments is typically associated with a particular individual. However, Web 1.0 hyperlink networks may be populated by nodes that are not homogeneous in type. For example, it is very easy to construct a hyperlink network where nodes will represent organisations that have an offline or real-world presence such as universities, government departments, companies or NGOs. But the network may also contain nodes that represent entities with no offline presence, that is, which only exist on the web. Examples are websites for online businesses, portals (sites that provide organised lists of links to other sites), information sites (sites that provide commentary on a particular topic and links to relevant resources), vanity websites (websites that have been set up to promote a brand or movie) and blogs. While in some research contexts it may make sense to prune the hyperlink network so that it only contains nodes representing organisations that exist offline, in others it may make sense to include all websites found by the web crawler, regardless of their type.

Even if a decision is made just to focus on websites that represent organisations with an offline presence, there are additional methodological challenges that are particular to hyperlink network analysis. Ideally, the nodes in an organisational hyperlink network will represent the entire Web 1.0 presence of each organisation, that is, websites (or parts of websites) rather than individual web pages. One way to achieve this is to group together all pages from a particular hostname. However, sometimes an organisation's web presence is reflected in multiple hostnames – different domain names reflecting, for example, international business presence (www.robscompany.com, www.robscompany.com.au) or different subdomains (brand1.robscompany.com, brand2.robscompany.com) or different subsites (e.g. www.robscompany.com/brand1, www.robscompany.com/brand2).[3] In order to accurately measure the web presence of the entire organisation all the identified pages from these domains, subdomains and subsites will need to be grouped together.

Hyperlink network ties

The second methodological question for hyperlink network analysis is: what are the network ties? Even in the simplest case of two organisations whose web presence is reflected in single websites, there are several ways we can construct a network tie. If the website of organisation i contains a hyperlink to the website of organisation j, then this could be a directed edge in the hyperlink network. But you might be more interested in only recording a network tie if the hyperlinks are reciprocated (i links to j, and j links to i), leading to an undirected hyperlink network. Finally, you might want to attach values or weights to the network ties. For example the total number of hyperlinks directed from i to j might be regarded as reflecting the strength of the connection. Alternatively, the depth in the website where the hyperlink was embedded might be used as the weight, since one could argue that a hyperlink from the homepage has more significance than if it is buried deep within the website.

Hyperlink network boundaries

The final methodological question is: what are the boundaries to the hyperlink network? In an offline friendship network the boundary might be the school (or classroom): if a student has a friendship with someone outside the school, then this friendship will not be included in the network. While geographical boundaries will often not make sense in the borderless terrain of the web, in some situations, geography might be helpful in determining hyperlink network boundaries. Lusher and Ackland (2011), for example, only included the websites of Australian-based organisations involved in refugee and asylum-seeker advocacy. However, even in a situation where there is an obvious network boundary, the analyst may still be faced with the problem of discovering all of the websites that exist within the boundaries and (if necessary) drawing an appropriate sample of websites. Lusher and Ackland started with an initial list of known relevant sites and then used the

---

[3]Note that subsites can cause a further problem: it may be that two or more organisations in your hyperlink network have websites that are commercially hosted, and in such a situation one needs to be careful that these websites are not merged into a single node reflecting their shared hostname.

VOSON web crawler (Section 4.3.1) to find additional relevant sites that were linked to the initial sites – a form of snowball sampling (Section 2.2.1).

4.2 THREE DISCIPLINARY PERSPECTIVES ON HYPERLINK NETWORKS

In this section, we introduce three disciplinary perspectives on the quantitative study of hyperlink networks.[4]

4.2.1 Citation hyperlink networks

Webometrics is a collection of techniques for quantitatively measuring documents and information from the web, and has its disciplinary origins in informetrics, which is a subfield of information science (Section 1.4). Thelwall (2009b) notes that webometrics has four main areas (text content analysis, analysis of hyperlink structure, web usage analysis and web technology analysis), but here we focus on webometric contributions to hyperlink analysis.

Section 9.2.1 presents an example of webometric hyperlink research, the study by Barjak and Thelwall (2008) of the factors associated with the prominence or visibility of websites belonging to research teams in the biological sciences. In this study, counts of inbound links are analysed along with the characteristics of the site and site owner, in order to identify those qualities that influence inbound links. The study aims to assess the viability of hyperlinks as scientometric performance indicators, and it highlights the fact that webometric hyperlink research involves the conceptualisation of inbound hyperlinks as being akin to citations that are traditionally studied by informetricians.

We refer to the technique used in Barjak and Thelwall (2008) as *hyperlink-counts regression* because it involves statistical regression analysis to find the relationship between a *dependent variable* (inbound hyperlink counts) and several *independent variables* reflecting characteristics of the site and the site owner.

It should be noted that webometric techniques have been used in areas outside of scientometric research. For example, Margolis et al. (1999) and Gibson et al. (2003) focus on counts of inbound hyperlinks to minor and major political party websites in their test of the hypothesis that the domination of major political parties offline is reflected on the web (Box 7.1). The use of counts of inbound hyperlinks to and outbound hyperlinks from government department websites in establishing the *nodality* of different government department websites (Escher et al., 2006) is also an application of webometrics (Section 8.1).

---

[4]This section draws on Borquez and Ackland (2012).

4.2.2 Issue hyperlink networks

The concept of issue networks (see, for example, Rogers, 2010a, b) is an example of studying hyperlinks from the media studies (Section 1.4) perspective.[5] Issue networks emerge when actors who are engaged in a common issue generate an *associational space* ('you are what you link to'), which is defined by hyperlinks. Another hallmark of this research is that it engages with public sphere theory (e.g. Habermas, 1989) in that it regards the web as a 'debate space'.[6] According to Rogers (2010b), issue networks are the representation of public debate (the initial concept was a circle diagram depicting actors sitting at a virtual table to discuss a particular issue), where the network nodes can be people and organisations, but also 'argument objects' such as news items, documents, or any type of content that is relevant to the issue. The conceptualisation of issue networks draws on actor network theory (Latour, 2005), envisaging actors that are both material and conceptual.

4.2.3 Social hyperlink networks

The potential for using social network analysis (Chapter 3) to analyse hyperlink networks was noted by Jackson (1997), who considered that SNA 'has significant potential to generate insight into the communicative nature of Web structures'. But Jackson (1997) was not comfortable with either the idea of nodes in a hyperlink network (pages or sites) being described as social actors or the core assumption of SNA – the interdependence of nodes within a network – as being applicable to the web.

While Jackson (1997) was not sanguine that formal SNA concepts and methods could carry over to the web, other authors have had fewer reservations, and Park (2003) advocated that the analysis of hyperlink networks using SNA be called 'hyperlink network analysis'. However, despite this early recognition of the potential of SNA for hyperlink analysis, there are not many examples of formal SNA techniques being used to analyse hyperlink networks.

Shumate and Dewitt (2008), Gonzalez-Bailon (2009) and Ackland and O'Neil (2011) (see Chapter 6) use ERGM in the context of analysing organisational collective behaviour on the web. Lusher and Ackland (2011) refer to the application of ERGM to hyperlink networks as *relational hyperlink analysis* and show that this approach can provide fundamentally different conclusions about the social processes underpinning hyperlinking behaviour, compared to hyperlink-counts regressions. In particular, hyperlink-counts regressions may overestimate the role of actor attributes in the formation of hyperlinks when endogenous, purely structural network effects are not taken into account.

---

[5] See also http://www.digitalmethods.net

[6] Public spheres are 'the social sites or arenas where meanings are articulated, distributed, and negotiated, as well as the collective body constituted by, and in, this process' (http://en.wikipedia.org/wiki/Public_sphere).

4.2.4 Comparing the disciplinary perspectives

Having introduced the three disciplinary approaches to researching hyperlink networks (citation hyperlink networks, issue hyperlink networks and social hyperlink networks), an attempt is now made to show how they differ. The three interrelated questions fundamental to the definition of a network (Section 3.1) – who are the nodes, what are the edges, and where are the boundaries? – can be used to show differences between the three approaches to hyperlink network research. Other useful dimensions on which to compare the three approaches to conceptualising hyperlink networks are network type (Section 3.1) and the unit of analysis.

Who or what are the nodes?

For citation hyperlink networks and social hyperlink networks, the nodes are typically websites that represent groups or organisations (which may or may not have an offline presence) or individuals (e.g. blogs). With issue hyperlink networks, as noted above, the nodes can represent either social actors or objects (e.g. news items, documents) that are relevant to the issue being studied.

What constitutes a tie?

At one level, this question seems quite simple because a hyperlink has a clear definition in terms of the technologies (HTML and HTTP) that allow one to connect documents across the web. But what meaning do we ascribe to the existence of a hyperlink?

As with the question above about who the nodes are, there are similarities in what constitutes a tie in citation hyperlink networks and social hyperlink networks. Typically, hyperlinks in such networks indicate positive rather than negative affect relations. That is, the hyperlink is likely to be transferring symbolic or practical resources that the recipient will want to receive. For example, in Barjak and Thelwall (2008) hyperlinks to academic project websites are conveying practical resources in the form of intellectual authority, status or prominence (Section 9.2.1). In Ackland and O'Neil (2011), the hyperlinks between environmental activist websites are modelled as conveying both practical resources ('index authority') and symbolic resources ('boundaries of belonging'), which help to establish online collective identity (Section 6.3).

As noted by Lusher and Ackland (2011), there is a point of departure between citation hyperlink networks and social hyperlink networks in terms of the nature of the tie. Webometrics is generally focused on the characteristics of actors that lead to the *acquisition* of hyperlinks (actor–relation effects, using ERGM terminology – see Section 3.2.2). Even though they used regression analysis rather than ERGMs, Barjak and Thelwall (2008) are effectively identifying a particular type of actor–relation effect, receiver effects, which are the website attributes that are associated with a tendency to receive hyperlinks. In contrast, social hyperlink network research involves identifying all possible social forces that have led to the emergence of a given hyperlink network. ERGMs are particularly well suited to this task, since they can be used to identify both actor–relation effects and hyperlinks that are unrelated to actor attributes, representing more informal networking that occurs between social actors because of social norms such as reciprocity (i.e. endogenous or structural network effects).

Once again, when it comes to the nature of the tie, the issue network is quite different from the social hyperlink network and citation hyperlink network. An issue network contains any actor with a stake or involvement in an issue, and for this reason we can expect the existence of negative affect relations in issue networks. An environmental issue network might contain environmental activists, government and business actors, and it is likely that hyperlinks from activists to business (and possibly government) will represent negative affect relations. In contrast, the environmental activist networks studied by Ackland and O'Neil (2011) only contained activists. In their study of refugee and asylum-seeker advocacy networks, Lusher and Ackland (2011) purposely excluded government websites because their presence would make it harder to interpret the meaning of hyperlinks owing to the fact that there would be a mix of negative and positive affect relations.[7]

Finally, because issue networks can contain nodes that represent objects (news items, documents) that pertain to an issue, it is difficult to conceive of an issue network as containing either actor–relation or purely structural effects. This is another difference between issue networks and the other two types of hyperlink networks.

Where is the network boundary?

In the case of citation hyperlink networks, one would think that the network boundaries will generally be clear, in the sense that the researcher will have a list of entities that exist in the real world (e.g. research teams, university departments, universities) and these entities comprise the actors of the network. Either all the entities will be included in the analysis (i.e. a census) or else a sample of the entities will be included. But it needs to be realised that citation hyperlink network analysis generally involves the construction of counts of inbound links from *anywhere on the web*, not just from actors included in the analysis. So in that sense, in the case of citation hyperlink network analysis the network boundaries are not as clear cut as one might first think.

In the case of social hyperlink networks, in some situations the network boundaries might be clear. For example, a study of the hyperlink networks

---

[7]The assumption here is that research is being conducted in an unobtrusive manner (digital trace data) and at a scale such that it is not feasible for the researcher to go to each page where there is a hyperlink and ascertain the context or meaning of the hyperlink.

of environmental activists might focus only on those organisations that have an offline presence, that is, activist organisations that are registered (for tax purposes) as non-profit organisations working on the environment. In such a situation the researchers could decide whether to conduct a census (include all such organisations in the study) or draw a random sample of organisations. However, with many examples of social hyperlink network research the underlying population of relevant websites cannot be identified in advance since the relevant population might be, for example, all websites run by organisations that are focused on a particular issue. In such a case, a snowball sampling approach is needed to draw additional nodes into the network.

With an issue network, as with many examples of social hyperlink networks, the underlying population of relevant websites cannot be identified in advance and a snowball sampling approach is again needed to draw additional nodes into the network.

Network types

We can use network types (Section 3.1) to further distinguish citation, social and issue hyperlink networks. A citation hyperlink network is in fact a series of 1.0-degree ego networks, with the focal nodes being the sites of interest. In contrast, social and issue hyperlink networks will generally be complete networks.

What is the unit of analysis?

The main technique used to analyse citation hyperlink networks is regression analysis, and hence the unit of analysis is the website. In contrast, social hyperlink networks are analysed using ERGMs and thus the unit of analysis is the dyad (pairs of connected websites). With issue networks, the unit of analysis is any object or actor that is connected to the issue.

4.3 TOOLS FOR HYPERLINK NETWORK RESEARCH

This section provides an introduction to web crawlers and also discusses sources for historical web data. Approaches for collecting data from blogs are also discussed.

4.3.1 Web crawlers

A web page contains two types of data that are relevant to this book, hyperlinks and text content, which are embedded in the HTML content in the page. While it is possible to use a web browser to collect these data from web pages, this approach is very time-consuming and not feasible for large numbers of websites (or indeed for websites that contain a lot of pages). For some time now researchers have been using *web crawlers*, which are software tools that automatically traverse a website by first retrieving a single web page (e.g. the entry or top-level page on a site) and then recursively retrieving all web pages in the site by following internal hyperlinks. A web crawler can save a copy of each web page that it encounters, but it can also parse the HTML content, extracting the hyperlinks and text content. In order to automatically extract data from web pages, web crawlers rely on the web pages being written to comply with the HTML specification (Box 4.1).

Three web crawlers used in the social sciences

The following are three examples of publicly available web crawlers that are used for social science research.

IssueCrawler[8] is web-based software for hyperlink network construction and analysis, developed under the direction of Richard Rogers. The first version of IssueCrawler was released in 2001; it is a pioneering example of a research tool that is accessible via a web browser. (Such tools are increasingly common, with the movement of services such as email and productivity tools into the 'cloud'.) With its origins in media studies (although widely used in the social sciences), IssueCrawler has been primarily designed as a web crawler for the construction and visualisation of issue hyperlink networks (Section 4.2.2).

Starting with a list of *seed URLs* (e.g. web pages focused on a particular issue), IssueCrawler crawls the seed set using three different approaches (Govcom.org, 1995). First, 'co-link analysis' is where only those websites receiving at least two links from the seed set are included in the final network. Of note is the fact that a seed site will be excluded from the final network if it does not receive links from at least two other seed sites. Co-link analysis in IssueCrawler needs to be differentiated from how the term co-link is used in information science, where it refers to the construction of a tie between two entities (e.g. articles, authors) because they both link to a third entity (even if they do not directly link to one another). In IssueCrawler, co-link analysis simply refers to the method by which websites are selected to appear in the final network, with links between websites in the final network being hyperlinks.

---

[8]http://govcom.org, http://www.issuecrawler.net

BOX 4.1   HTML AND RDF

The following is the HTML for the homepage of a fictitious travel website (http://www.robstravel.com):

```
<!DOCTYPE HTML PUBLIC "-//W3C//DTD HTML 4.01//EN" "http://www.
w3.org/TR/html4/strict.dtd">
<HTML>
   <HEAD>
      <TITLE>My first HTML document</TITLE>
      <META name="keywords"content="travelservice,holidays,Australia">
      <META name="title" content="Rob's Travel">
      <META name="description" content="We specialise in holidays to
      Australia!">
   </HEAD>
   <BODY>
      <P>Welcome to Rob's Travel!</P>
      <P>You may also like to see the website of our partner
      <A href="http://www.robsrentalcars.com">Rob's Rental Cars</A>,
      who rent cars.</P>
   </BODY>
</HTML>
```

The meta keywords and description provide information about the nature of the site, while the page body contains a hyperlink to a partner website (http://www.robsrentalcars.com). A web crawler can extract the metadata, body text and hyperlinks, but social science hyperlink research will often require manual coding of the site and possibly even hyperlinks. It has been suggested that the Semantic Web might obviate the need for manual coding of sites and hyperlinks, since the framework involves the use of markup languages that allow website owners to present basic information about the site in a machine-readable manner. To illustrate, the above example could be 'semantified' with the addition of the following RDF document:

```
<?xml version="1.0"?>
<rdf:RDF xmlns:rdf="http://www.w3.org/1999/02/22-rdf-syntax-ns#"
xmlns:si="http://www.robstravel.com/rdf/">
<rdf:Description rdf:about="http://www.robstravel.com">
   <si:title>Rob's Travel</si:title>
   <si:product>travel service</si:product>
   <si:category>holiday</si:category>
   <si:country_specialisation>Australia</si:country_specialisation>
   <si:partner>Rob's Rental Cars</si:partner>
   <si:partner_url>http://www.robsrentalcars.com</si:partner_url>
   <si:partner_product>rental cars</si:partner_product>
</rdf:Description>
</rdf:RDF>
```

The Rob's Travel semantically enabled website would use an *ontology* (an exact description about objects and their relationships) providing precise meanings of the tags 'product', 'category', etc. As long as the web crawler understands the ontology, then the site could be automatically coded as a business that is providing travel services, specialising in holidays to Australia, which is affiliated with a rental car company.

However, Brent (2009) contends that the Semantic Web may not be a boon for social science web researchers for two main reasons. First, the human effort involved in making a website both human- and machine-readable (i.e. presenting the information in both HTML and RDF) is significant, and while there might be incentives for this to happen in particular domain areas (e.g. libraries and e-commerce), it is less likely to occur in areas of interest to social scientists (e.g. advocacy and protest sites). Second, it is questionable that a single ontology can adequately capture the diversity of parts of the web that are of interest to social scientists. While one can envisage real-world examples of ontologies successfully being used to make e-commerce sites machine-readable, it is hard to imagine organisations participating in, for example, the abortion debate agreeing to and then implementing an ontology that would allow social scientists to automatically code their websites on the basis of their stance on abortion.

The second technique for constructing a network in IssueCrawler is 'snowball analysis', which is equivalent to snowball sampling (Section 2.2.2): all of the outbound links from the seed sites are included in the network and these sites are then themselves crawled (this process continues up to three degrees of separation from the seed set). Finally, 'inter-actor analysis' displays interlinking between the seed set exclusively, that is, it allows the construction of a complete network.

SocSciBot[9] is a long-established web crawler and hyperlink network analysis tool that was developed by Mike Thelwall (see Thelwall, 2009b). Unlike IssueCrawler, SocSciBot is client software (you download and install it on your own computer). SocSciBot was developed for the construction and analysis of citation hyperlink networks (Section 4.2.1); the earlier versions of SocSciBot were primarily designed for measuring web impact through retrieving counts of inbound links and did not include a network visualisation tool. However, as with IssueCrawler, SocSciBot can be used for other types of hyperlink research. For example, Shumate and Dewitt (2008), which we regard as an example of social hyperlink network research, employed SocSciBot for data collection, and

---

[9]Statistical Cybermetrics Research Group, University of Wolverhampton, http://socscibot. wlv.ac.uk

the current version of SocSciBot provides network visualisation and is designed to be used in combination with the social network analysis tool Pajek.[10]

Virtual Observatory for the Study of Online Networks (VOSON)[11] is a tool for collecting and analysing hyperlink network data and, like IssueCrawler, is a hosted service available via a web browser. VOSON was first publicly released in 2006 and has been designed specifically for the analysis of social hyperlink networks (e.g. Lusher and Ackland, 2011; Ackland and O'Neil, 2011). A VOSON plug-in to the NodeXL social media analysis tool is available (see Ackland, 2010b).

Application programming interfaces

SocSciBot, IssueCrawler and VOSON all feature web crawlers as their main data collection tool. However, often we want to know not just the hyperlinks that are being directed *from* a given website (i.e. outbound hyperlinks), but also what hyperlinks are being directed *to* the website (i.e. inbound hyperlinks). Search engines such as Google and Bing allow you to find this information manually via their search engine websites: put "link:voson.anu.edu.au" into Google, and it will list all of the web pages that hyperlink to the VOSON website. However, for large-scale research, the Google and Bing search sites are not useful, and instead it is better to use application programming interfaces (APIs) to enable software to query the databases directly.[12] VOSON uses various APIs to find inbound hyperlinks to seed URLs, and *Webometric Analyst*[13] (formerly called LexiURL Searcher) is a companion tool to SocSciBot that also enables the collection of inbound hyperlinks. For more on APIs in the context of hyperlink research, see Thelwall (2004, 2009b).

Ethics of using web crawlers

In addition to the points made in Section 2.8, there are particular ethical issues associated with the use of web crawlers for research (Thelwall and Stuart, 2006). First, crawling a website can potentially use a lot of the resources (e.g. bandwidth, CPU time) of the website owner, which could lead to significant costs or a loss of service quality. For this reason it

[10]http://vlado.fmf.uni-lj.si/pub/networks/pajek

[11]VOSON Project, the Australian National University, http://voson.anu.edu.au. It should be noted that the author created the VOSON software and is involved in its commercial development.

[12]See http://code.google.com and http://www.bing.com/toolbox/bingdeveloper

[13]http://lexiurl.wlv.ac.uk

is important that web crawlers are used responsibly, for example by not crawling the sites of organisations that might be resource-constrained (e.g. NGOs in developing countries), and also by limiting the crawler so there are delays between each page request. Second, it is important that web crawlers obey the robots.txt protocol,[14] which is used by webmasters to inform crawlers which parts of the website can be crawled and which parts are 'off limits'.

4.3.2 Historical web data

While the above tools can be used to collect 'live' website data, what if we want to study how a group of websites have changed over time?[15] For example, suppose we were interested in using web data to look at how climate change has emerged as an issue of concern, and how various actors (governments, corporates, NGOs) are responding to climate change. To do this we either need access to historical web data or else we need to have been periodically crawling the web ourselves, thus constructing a time series of webcrawls (and network datasets).

The Internet Archive[16] was founded in 1996 as an Internet library, offering 'permanent access for researchers, historians, and scholars to historical collections that exist in digital format', and it is a member of the International Internet Preservation Consortium (Box 4.2) which also includes many national libraries as members. The Internet Archive has been crawling the web since 1996, and currently the archived web pages are publicly available via the Wayback Machine, a browser interface. There is currently no way to automatically extract hyperlinks or text content from websites that have been archived by the Internet Archive. This means that while the Wayback Machine is useful for qualitative or descriptive research involving a single website or a small number of websites, it is not suitable for large-scale empirical research. It would be impossible, for example, to study the evolution of the hyperlink networks formed by a couple of hundred environmental activist websites. However, the Internet Archive has plans to develop an API that will hopefully facilitate automated querying of their historical hyperlink data in a similar way to the Google and Bing APIs.

---

[14]http://www.robotstxt.org

[15]Note that data collected via the Bing and Google APIs will not necessarily be current, since the data are extracted from the databases of these organisations; however, they are continually crawling the web so we can expect the data will be reasonably up to date.

[16]http://www.archive.org

4.3.3 Blogs

The underlying technology of a blog is the same as that of a website: a blog
is simply a chronologically updated website, generally authored by a single
person, with a diary or commentary style. Web crawlers (such as from blogs

SocSciBot andVOSON) can extract hyperlinks and text. However, the structure of blog pages makes the data collection via web crawlers more challenging than is the case with static websites, for two reasons.

First, standard web crawlers will not be able to distinguish between *permalinks* (hyperlinks that appear within a given blog post), *blogroll* links (the hyperlinks that often sit to the side of the page, i.e. are not contained within a particular blog post), and links that might appear in the comments section on the blog page (i.e. links not made by the blogger, but by someone else commenting on the blog post). Permalinks are generally made when the blogger comments on or points to other blog posts or web pages on traditional media sites. Permalinks thus generally reflect the current reading behaviour of the blogger, and are often considered by researchers to be a more accurate indicator of the links that bloggers are making to one another and to other sites. In contrast, blogroll links generally reflect more permanent affiliations between bloggers. For example, it is common for political bloggers to have blogroll links to other bloggers who share the same political persuasion. Blogroll links can become 'stale' and may therefore be of less value in blog analysis.

The second challenge faced in collecting data from blogs relates to the chronological structure of the average blog page. Blog sites are generally structured so that all the blog posts for a given month appear on the same page (with the most recent posts at the top of the page). One of the main aims of blog research is to identify links that are made within a given time period. This allows the tracking of the diffusion of influence throughout the blogosphere, for example (where was an issue first taken up, and how did it spread throughout the blogosphere?). In order to get this type of dynamic data, it is necessary to ensure that the blog pages are parsed so that links can be attributed to a particular time period.

The above challenges regarding the collection of blog data are not insurmountable, but researchers are advised to make use of specialised services for providing data from the blogosphere. An example of such a service is the Blog Analysis Toolkit,[17] which is a hosted web crawler specifically designed for blogs, while Infoscape[18] has also developed tools for extracting data from the blogosphere. In addition there are several APIs for blog data: Ackland (2005) used Bloglines,[19] which was an early blog API, and Spinn3r[20] provides a blog API and also has released several large blog datasets to the research community (Burton et al., 2009).

[17]https://surveyweb2.ucsur.pitt.edu/qblog/page_login.php

[18]http://www.infoscapelab.ca

[19]http://www.bloglines.com

[20]http://www.spinn3r.com

4.4 CONCLUSION

This chapter began with a discussion of various motives for creating hyperlinks and also the benefits an organisation may gain from receiving a hyperlink. It then discussed the definition of a hyperlink network, focusing on how nodes, ties and boundaries to networks can be understood in the context of hyperlink networks. The chapter then proposed three broad disciplinary approaches for empirically studying hyperlinks from a social science perspective, where hyperlink networks can be conceived of as citation, issue or social networks. The final section was devoted to tools for collecting and analysing hyperlink data, focusing on three web crawlers that are used for hyperlink research: IssueCrawler, SocSciBot and VOSON.

Further reading

For more on webometrics and the use of web crawlers in the social sciences more generally, see Thelwall (2004, 2009b) and also the SocSciBot website.[21] The Digital Methods Initiative website[22] and Govcom.org website[23] provide more on the concepts and methods relating to issue networks and the IssueCrawler software. Finally, more information about the VOSON software and related methods can be found on the VOSON project website.[24]

---

[21]http://socscibot.wlv.ac.uk

[22]http://wiki.digitalmethods.net/Dmi/WebHome

[23]http://govcom.org

[24]http://voson.anu.edu.au

PART II

WEB SOCIAL SCIENCE EXAMPLES

5

Friendship Formation and Social Influence

This chapter looks at why people form friendships and other social ties, and the role of social networks in the formation of people's preferences (towards music, food, movies, etc.). Web data are providing new insights into these long-standing research topics in social science.

Section 5.1 provides an introduction to research into homophily, or the tendency that 'birds of a feather flock together'. The section looks at empirical challenges with identifying homophily and examples of research into homophily using data from Facebook and online dating sites. Section 5.2 looks at the other side of the coin, introducing research into social influence or the phenomenon where people become more like their friends. The section discusses the empirical challenges in identifying social influence and then provides examples of social media research into social influence.

5.1 HOMOPHILY IN FRIENDSHIP FORMATION

There is strong evidence that people 'assortatively mix' when it comes to forming friendships and other social connections such as sexual partnerships and marriage. Sociologists have found that patterns of friendship are strongly affected by characteristics such as age, race and language – the 'birds of a feather flock together' phenomenon. With regard to marriage, research reviewed in McPherson et al. (2001) shows that Americans exhibit a preference for 'same-race alters' far in excess of preference for similarity based on other characteristics such as age and education. This section defines assortative mixing and homophily, and explores how homophily can be empirically measured. We also look at how web data are being used for research into homophily.

5.1.1 Measurement issues

Assortative mixing refers to a positive correlation in the personal attributes or characteristics (age, race, ethnicity, education, religion, socio-economic status, physical appearance, etc.) of people who are socially connected to one another. Assortative mixing is an empirical measure that simply describes the structure or composition of a social network. It shows which types of nodes have a higher probability of being connected to one another. Assortative mixing is therefore an observable network pattern, but it says nothing about the exact processes that have led to the formation of a particular social network. While it is reasonably easy to measure the level of assortative mixing in a social network, it is more difficult to discover *why* people are assortatively mixing. There are three main reasons why a given social network might exhibit assortative mixing.[1]

First, there might be *homophily* – a term first coined by Lazarsfeld and Merton (1954) which refers to people forming a social tie purely because they share a particular attribute (they want to be connected to someone who is similar). Homophily can in principle operate with respect to any attribute – physical characteristics such as race and sex, and cultural preferences over books and music. However, as we will see in Section 5.2, when the attribute is changeable (the person can choose the attribute) it becomes harder to distinguish whether 'birds of the feather are flocking together' (attributes are influencing friendship formation) or whether someone is becoming more like their friends (friendships are influencing attitudes and preferences).

Second, there may be *opportunity structures* that influence social tie formation. In particular, *group size* is important: the smaller a particular group (e.g. racial category) the more likely (all other things considered) that its group members will form social ties outside of the group. If group size is not controlled for, then there can be erroneous conclusions about the homophilous behaviour of different groups who have different population shares. Independent of group size, the *propinquity mechanism* can also influence whether two people form a social tie – this might relate to spatial proximity (e.g. living in the same neighbourhood) or shared institutional environments (e.g. working in the same organisation).

Finally, there are other types of network substructures or mechanisms that (unlike homophily) are not related to the attributes of actors in the dyad but still might influence social tie formation and create higher levels of assortative mixing in the network. For example, there is the property of *sociality*: two people from, for example, the same ethnic group might become friends simply because they are both social people and like to make lots of friends. This would be contributing to the level of assortative mixing in the network but it is not because of homophily, but another actor–relation effect known as the sender effect (Section 3.2.2).

As we saw in Section 3.1, social networks tend to exhibit two properties: *reciprocity* – if i extends the hand of friendship to j, there is a good chance that j will reciprocate the friendship – and *transitivity* – the tendency for friends-of-friends to become friends. These two balance mechanisms (reciprocity and transitivity) can also affect the measurement of

---

[1]See Wimmer and Lewis (2010) for a detailed review.

homophily: if a particular group does have a genuine preference for form-ing in-group ties (i.e. homophily), then this preference will be amplified by the processes of reciprocity and transitivity. Furthermore, if there are differences in these balance mechanisms across different social groups (e.g. one race has a cultural tendency to reciprocate friendships or introduce friends to each other) then this will tend to obscure the cross-group com-parison of homophily.

The problem for researchers studying homophily is that both opportunity structures and network effects other than homophily can 'mask' the true level of homophily in a social network. However, as we saw in Section 3.2.2, homophily is a particular type of actor–relation effect that can be modelled in ERGMs. Thus it is possible to use ERGMs to establish whether there is significant homophily in a network after controlling for other factors (e.g. group size, endogenous network effects). However, it is also possible to con-trol for group size using a simple non-statistical procedure (Box 5.1).

5.1.2 Friendship formation in Facebook

Wimmer and Lewis (2010) used data from Facebook to investigate the role of race in friendship formation. They collected Facebook data for a cohort of an 'East Coast private college'; see Lewis et al. (2008) for more details on the dataset (see also Section 3.3.3). To construct the friendship network, they used the pictures of friends posted by students on their personal pages and they only used the subset of the population posting pictures of their friends (a sample of 736 students).

Wimmer and Lewis (2010) used their novel dataset and ERGM to provide several new insights into friendship formation. First, they found that the use of over-aggregated racial categories in previous research may have masked ethnic in-group preference. The authors found that some racial categories matter for social network formation (especially the 'black' category), but not others (e.g. the 'Asian' category). A finding of apparent homophily among 'Asians', for example, is largely an artefact of South Asians sticking together with other South Asians, Chinese with other Chinese, etc. Similarly, some ethnic categories (e.g. South Asian, Jewish, Chinese, British) matter a lot, but not others (e.g. Italian).

Second, they found that a failure to take into account opportunity struc-tures (in particular, the population share of different ethno-racial groups) and differences in the sociality of the members of these groups led to previ-ous studies over-exaggerating the homophily rates of large groups (usually whites) and those of groups comprised of more sociable individuals with larger networks.

Third, previous research into homophily used statistical techniques that ignored triadic closure effects, thus leading to an overestimation of the ten-dency towards homophily, including racial homophily. A black person might become friends with another black person not so much because each prefers same-race alters, but because they share a friend in common.

BOX 5.1   HOMOPHILY IN A US SCHOOL FRIENDSHIP NETWORK

Much of the research into friendship homophily has focused on school students. There are important social and psychological implications of friendship homophily in schools, and promoting interracial friendships at an early age might contribute to a more tolerant and inclusive society. However, there are also pragmatic reasons why homophily research has focused on schools. These are good places to collect data on friendship networks since school children often have most of their friends in the same school (or even the same class) and hence it is much easier to collect complete network data.

Currarini et al. (2009) calculate homophily measures using data from the Adolescent Health dataset, which has information on friendship patterns in a representative sample of US high schools. Let s_i denote the average number of friends that a person from group i has with people from the same group, and let d_i denote the average number of friends that a person from group i has with people from other groups. The basic assortative mixing or *homogeneity* index H_i measures the fraction of ties that individuals of type i have with that same type: $H_i = s_i \div (s_i + d_i)$. Note that Currarini et al. (2009) refer to this as the homophily index; however, we follow Wimmer and Lewis (2010) by using the term 'homogeneity', which is synonymous with assortative mixing (defined above).

The problem with the index H_i is that it is partly determined by the relative population sizes of the different groups. Take the example where one group accounts for 95% of the population and has 96% of its ties 'within-group', and another group accounts for 5% of the population and has 96% of its ties with the same type. Both groups have the same homogeneity index (0.96), but they are very different in terms of how homophilous they *are*, compared to how homophilous they *could be*. We want a measure of homophily that (in effect) controls for group size.

Define group i's proportion of the population as w_i. The homophily index H_i^* of group i, $H_i^* = (H_i - w_i) \div (1 - w_i)$, normalises the homogeneity index by the potential extent to which a group could be biased towards its own type. If there is pure baseline homophily (the percentage of in-group ties the group forms is equal to its share of the population, $H_i = W_i$) then $H_i^* = 0$. 'Inbreeding' homophily (using the terminology of Coleman, 1958) is indicated by $H_i^* > 0$, and if there is complete inbreeding ($H_i = 1$) then $H_i^* = 1$. Finally, inbreeding *heterophily* is indicated by $H_i^* < 0$.

In one of the schools reported in Currarini et al. (2009), white students account for 51% of the student population, while black students comprise 38% and Hispanics 5%. For white students, 85% of their ties are within-group (i.e. the homogeneity index for whites is 0.85). So, H_{whites}^* is $(0.85 - 0.51) \div (1 - 0.51) = 0.69$ indicating that there is some inbreeding homophily within this group. In contrast, with hispanics, the homogeneity index is 0.02 and hence there is some evidence of inbreeding heterophily for this group: $H_{Hispanics}^* = (0.02 - 0.05) \div (1 - 0.05) = -0.032$

Finally, the datasets that have been used for previous research into homophily only contained the most basic demographic attribute data. The richness of the Facebook data allowed Wimmer and Lewis (2010) to find that other characteristics are just as important, if not more important, for social network formation.

5.1.3 Online dating

A lot of previous research into assortative mixing in sexual partnerships has used marriage data. The problem with using marriage data in this context is that marriage is an equilibrium outcome which arises from the process of search, meeting and choice. If the aim is to establish the exact reasons for assortative mating then marriage data are clearly problematic. This is one of the reasons why researchers in this field are turning to online dating sites as sources of detailed information on personal attributes and behaviour (showing and receiving interest).

Online dating: background

Online dating has become one of the most successful online businesses. According to the Pew Internet & American Life project, in 2006, 37% of all US Internet users who were single and looking for a partner had gone to an online dating site (Madden and Lenhart, 2006). As Griscom (2002) put it: 'Twenty years from now, the idea that someone looking for love won't look for it online will be silly, akin to skipping the card catalog to instead wander the stacks because "the right books are found only by accident." ... But serendipity is the hallmark of inefficient markets, and the marketplace of love, like it or not, is becoming more efficient.' There are three types of online dating service:

- Profile-based (e.g. RSVP). Users register and create profiles, stating socio-economic and socio-demographic characteristics and personal preference such as likes and dislikes. Providing a photo is optional. Users can search profiles, and send and receive 'kisses' (individuals acknowledging each other using pre-defined messages) and emails. Eventually users might talk and meet offline.
- Algorithm-based (e.g. eHarmony). Users provide profile data which are then used to provide matches. Users do not get to search profiles.
- Social network (e.g. Friendster). Friendster was an early social network service (established before Facebook) and was originally set up for online dating.

Homophily in online dating

Fiore and Donath (2005) were granted access to profiles and logs from a US online dating site. The profile data included sex, age, height, location, physical build, smoking/drinking preferences, education, number of children, number of preferred children, and race. The dating service was predominantly used by whites (84% of users), and 55% of users were women, with a median age of 34 (males were slightly older).

During the data collection period (between June 2002 and February 2003), there were more than 110,000 exchanges of one or more messages between unique pairs of users (dyads). Men initiated 73% of conversations, and 21% of male-initiated conversations were reciprocated (compared with 25% of female-initiated conversations). The authors found that the number of ties per person follows a power law (Section 7.1.1), implying a very unequal distribution of attention or activity: a small number of people are involved in a large proportion of the recorded dyadic ties.

The authors investigated what characteristics are most 'bounding', that is, the dimensions on which users are more likely to seek someone like themselves. They constructed two measures: *actual sameness* refers to the percentage of contacts between two people sharing a value for a particular characteristic (e.g. 'athletic' for the characteristic 'physical build'), while *expected sameness* refers to the percentage that you would expect if males and females were paired randomly. Note that expected sameness for a particular attribute varies with the number of possible values for the attribute and the evenness of the distribution of users amongst the possible values of the attribute. Expected sameness is higher when the number of possible values is low, and when many users pick the same value for a particular attribute. Thus, in the case of this particular dating service where 84% of users were white, the expected sameness for race will be high.

The attributes for which there was the biggest difference between actual and expected sameness (i.e. the attributes with the highest 'boundedness') were marital status, number of children and preferences regarding children. The attributes with low boundedness were drinking habits, presence of pets and preferences regarding pets. The boundedness measure used by Fiore and Donath (2005) is therefore similar to the Currarini et al. (2009) homophily index (Box 5.1) in that group size is being controlled for.

What makes you click?

Hitsch et al. (2005) collected data from profiles and activity logs for more than 23,000 users of a US online dating service. The observations were collected for 3 months in 2003, and a subset of users were also manually rated for their attractiveness, based on their profile photo.

The authors conducted two main types of analysis. First, they wanted to see how the propensity to make contact is related to the attractiveness of the profile of the potential mate and the attractiveness of the searcher. If less attractive people are less likely to contact more attractive people (the authors are economists and explained it as net expected benefit of contacting an attractive person being less than net expected benefit of contacting a less attractive person), this would be evidence that the less attractive are 'relying on themselves'.[2]

The authors did not not find evidence that less attractive people are less likely to contact more attractive people. For men, the probability of sending a contact was found to increase more or less monotonically with the attractiveness of the recipient (female), and there was not a significant impact of the attractiveness of the sender (male). For females, the probability of sending a contact did not rise as markedly with increases in the attractiveness of the recipient (male), and it was found that (as with males), less attractive females are not less likely (compared with more attractive females) to make contact with more attractive males.

Hitsch et al. (2005) also looked at how attributes correlate with attention received in the dating site. They found that men who were rated in the top 5% based on attractiveness received on average over 300% more first contacts compared with the baseline (the average male), while men rated as being in the bottom 10% in terms of attractiveness received nearly 50% fewer first contacts compared with the baseline or control person. There was also a positive correlation with number of contacts received (more marked for men than for women) but with an apparent jump in interest for those people who used humour to describe their looks ('bring your bag in case mine tears'). Height was positively correlated with attractiveness of men, and negatively correlated for women. Finally, there was evidence that stated income attracts interest for men, but has no impact on the attractiveness of women. While they did not use the ERGM approach (Section 3.2.2), Hitsch et al. (2005) were therefore calculating something similar to the receiver effect, that is, how attributes affect the propensity to receive ties.

5.2 SOCIAL INFLUENCE

Understanding the processes by which social networks influence preferences and behaviour has been a major focus of social science research for decades. This section first presents a review of research into social networks and preference formation. It then looks at several examples of research on this topic where web data are used.

5.2.1 Identifying social influence

There are three possible reasons for an observed correlation in behaviour between individuals who are socially connected (see, for example, Van den Bulte and Lilien, 2001; Christakis and Fowler, 2007). First, there may be

---

[2]This is another reason for assortative mating which is different from the three factors discussed above (homophily, opportunity and endogenous network effects). This refers to the fact that there might be *agreed-upon preferences*: people tend to agree upon what characteristics are most attractive; the most attractive select among themselves and the less attractive must rely on themselves.

social influence.[3] Using teenager smoking as an example, social networks may influence smoking behaviour via the process of information transmission – students might update their beliefs about the costs and benefits of smoking after talking to another student who also smokes. It is also likely that normative pressures will play an important role – a non-smoking student who has friends who smoke may feel a sense of dissonance and therefore decide to smoke to fit in with his or her peers.

A second potential reason for socially connected individuals displaying similar behaviour is *social selection* or homophily (Section 5.1), which is the phenomenon that people who are similar are more likely to want to connect with each other (see, for example, McPherson et al., (2001). In the case of teenage smokers, this would be a case of smokers seeking out one another to become friends.

However, the situation is more complicated than just establishing whether it is a case of 'birds of a feather flocking together' (social selection), or people becoming more like their friends (social influence). A third reason why there may be correlation of behaviour within a social network is that individuals can share common unobservable (to the researcher) factors which influence their behaviour; we refer to such influences here as *contextual factors*. In the case of teenager smoking, the contextual factors might be, for example, marketing.

A hallmark of social network analysis is an emphasis on the influence of the structure and composition of a person's social network on his or her behaviour (and outcomes), over and above the role of the person's individual attributes (Section 3.2). Sociologists studying social networks have often taken it as a given that individual behaviour and outcomes are influenced by the structure and properties of social networks (although for an exception, see Mouw, 2006). In contrast, a lot of the work by economists in this area has focused on the need for model specification and identification in order to establish the presence and extent of social influence (the foundational reference is Manski, 1993).[4]

The following are some of the approaches that have been proposed for solving the problem of spurious correlation in the identification of social influence (for a summary see, for example, Soetevent, 2006; Reagans et al., 2007):

- *Selection effects may not exist.* In his analysis of social embeddedness in market interactions, Granovetter (1985) shows that in some cases social networks can be regarded as primordial, that is, arising prior to, and for reasons other than, the market transaction. The classic example he provides

---

[3]The term 'social influence' is typically used in sociology, while economists researching social interactions refer to 'peer effect', 'peer influence' or 'network effect'. More recently, the term 'social contagion' has been used.

[4]For example, Cohen-Cole and Fletcher (2008) use standard econometric techniques to assess the claim that obesity spreads through social networks (Christakis and Fowler, 2007), and show that after contextual factors are properly controlled for, there does not appear to be a social contagion effect for obesity.

Web Social Science Examples

is Hasidic diamond traders – the social networks are related to race rather than business, but are shown to have an impact on business behaviour and outcomes. In such cases, it is reasonable to assume that network selection effects are not present.

- *Natural or field experiments.* In some situations, naturally occurring interventions involving changes to groups allow researchers to accurately estimate social influence – these are called natural experiments (Section 2.1). Examples are Sacerdote (2001) who estimates peer effects among randomly assigned college roommates, and Katz et al. (2001) who study the influence on family well-being of a move from a high-poverty to a low-poverty area, by comparing families who received (randomly assigned) housing vouchers with families who did not receive the voucher. Field experiments (Section 2.1) are conducted outside of the laboratory but where the researcher still has control over the variable of interest. The Internet has provided some interesting opportunities for online field experiments aimed at extending understanding of social influence, as discussed further in Sections 5.2.2, 10.2.1 and 10.2.2.
- *Instrumental variables.* The instrumental-variables econometric technique can be used to control for spurious correlation; however, this technique is dependent on identifying a variable that is correlated with network formation but uncorrelated with the behaviour of interest. In practice, such instruments are often very difficult to find. See Box 7.5.
- *Panel data.* In their study of the impact of social networks on political attitude formation, Lazer et al. (2008) were able to collect data on the research subjects' political views before and after their exposure to one another. This allowed the authors to show how social interactions influence political views. Mercken et al. (2010) use a particular approach to longitudinal network analysis (actor-oriented models) to study the mutual influence of smoking behaviour and friendships in adolescence. Using data on Finnish students, they show that both selection and influence processes are important in explaining adolescent smoking behaviour, with the strength of both processes decreasing over time (also see Box 7.5).

5.2.2 Social influence in social media

The standard concept of social influence presented above is one where the behaviour of actors in a social network is changed or altered as a direct result of the social tie. The typical example we have discussed is smoking behaviour: a teenager is influenced by another friend and, as a result, starts smoking. This influence might be transmitted by word-of-mouth information (smoking is cool) or simply the teenager emulating what he or she sees. The change in behaviour is directly observable since the researcher knows whether or not the teenager smokes (but, as we have noted, identifying social influence as the cause of the change in behaviour is not straightforward).

The concept of social influence needs to be modified for social media environments in several key ways. First, as discussed previously, while it is possible to extract interpersonal networks from most social media environ-ments, some of these will be social networks while others will be more accurately described as information networks – or a subset of these, com-munication networks (Section 3.4). The concept of transmission of influence in a social network is different from that in an information network. Because the nature of the tie is very different, we can expect the one-to-one power of influence to be less in the sense of actor i being able to influence actor j to change his or her behaviour or thinking. We can think of a threshold effect here (Centola and Macy, 2007): influencing someone to go out and purchase a music video is one thing, but influencing someone to go out onto the street and protest is another. A tie in an information network may facili-tate the transfer of influence in the former but not in the latter (reinforcing ties, i.e. 'wide bridges', may be needed instead). But on the other hand, infor-mation networks are likely to be larger, so the cumulative influence (i.e. impact on society as a whole) might be greater. This is the difference between convincing a single person to undertake a radical and possibly dangerous act of protest (via a social network) and convincing thousands of people to go out on the streets (via an information network).

The second way in which we need to modify the concept of social influ-ence for social media environments relates to the observability of the out-come. Recall that the main focus of this book is on the use of unobtrusive methods for collecting digital trace data, with an emphasis on automatic methods of data collection. The methodological context necessarily restricts the range and type of outcomes that can be observed. While it is entirely possible that an individual may be influenced via interactions in social media to change his or her behaviour (whether it be attitudes to politics or con-sumer behaviour), we will not know this unless the behavioural change results in a digital trace that is observed by the researcher. The study by Aral et al. (2009) into social influence in an instant messaging network (see below) is focused on an outcome (product adoption) that is measured in the dataset. But what about changes in a person's political identity or willingness to vote? Unless this change is somehow identifiable in the data, or else we conduct a survey or experiment to gauge how attitudes have changed, this will never be known by the researcher.

This second issue above (observability of the outcome of influence) has greatly affected how influence in social media environments is conceptual-ised and studied. In particular, as is seen in the example below on Twitter, influence in social media environments is largely conceptualised as atten-tion. That is, what is really a dynamic process (i interacts with j and, as a result of the interaction, i's behaviour changes) gets reduced to a static ques-tion of who is the most central or visible or important person in the social media network, i.e. who is garnering the most attention. Wu et al. (2011) note that Lasswell's maxim that media communication studies is about 'who says what to whom in what channel with what effect' is difficult to apply in the case of Twitter (and most other social media environments) because it is very difficult to establish the effect of interactions in these environments.

But the other reason why *centrality* is often equated with *influence* is that there is an implication that there is a potential 'mapping' (Section 8.3) from online influence (measured by centrality) to offline influence. Thus while we may not be able to observe the direct effect of an influential person in Twitter (e.g. how that person influences another person's voting behaviour), the fact that the person is being listened to by a lot of people (garnering lots of attention) is a proxy for the strength of influence on a single person – the aggregate influence is what is important. Similarly, for webometric analysis of hyperlink structures, hyperlinks are being used as a proxy for offline academic productivity and impact (Section 9.2.1).

Product adoption in an instant messaging network

Aral et al. (2009) studied social influence in a global instant messaging (IM) network consisting of over 27 million users. Their dataset contained: day-by-day adoption data of a new mobile service application launched in July 2007 (Yahoo! Go); users' attribute data (demographics, geographic location, mobile device type); and users' cultural preference data (page views of different types of web content – news, sport, weather, finance, etc.).

The authors observed strong evidence of assortative mixing on the basis of adoption of Yahoo! Go service – that is, there was a strong correlation in this particular behavioural attribute (whether you have signed up for Go or not) amongst friends, with people who have signed up for Go tending to be IM friends with other Go adoptees. The fact that their data are dynamic (day-by-day adoption data) also allowed the authors to investigate the existence of temporal clustering, that is, the probability of someone who adopts having friends who also adopt within a short period of time (either just before or just after).

Together these two empirical facts in the data (assortative mixing and temporal clustering on the basis of adoption behaviour) are strong evidence of social influence, but, as Aral et al. (2009, p. 21545) note, there is another phenomenon that could be driving this phenomenon – homophily:

> Demographic, behavioral, and preference similarities could simultaneously drive friendship and adoption, creating assortative mixing [and temporal clustering]. ... If friends are more similar, they are more likely to have similar strengths of preference for Go and similar desires to be 'early adopters' of mobile technology services, making them more likely to adopt contemporaneously even if they do not influence one another.

In trying to measure the extent of social influence in this network, Aral et al. (2009) used a 'matched sampling' estimation approach (Brock and Durlauf, 2001). The idea behind this approach is to estimate the impact on the probability of a user having an adopter in his or her local (ego) network (this is the 'treatment'). But the problem is that adopters and non-adopters do not have an equal probability of being treated; that is, the treatment is not being randomly assigned, but is in fact being influenced by homophily. Adopters have a higher probability of being treated than non-adopters because homophily over characteristics related to adoption (e.g. being an 'early adopter' or tech-focused person) is driving friendship formation.

The matched sampling approach involves using logistic regression to estimate the probability of a person being treated (having one or more adopter friends) where the regressors are the demographic and behaviour characteristics noted above. The regression was then used to identify a group of people who, based on their characteristics, *should* have been treated (but in fact had not, i.e. did not have any adopter friends). This control group was then compared with a treated group, and this allowed the authors to estimate upper bounds on the degree to which social influence (rather than homophily) explains assortative mixing and temporal clustering of adopters in the network. The approach allowed the authors to conclude that previous methods overestimate social influence in product adoption decisions in this instant messaging network by between 300% and 700%, and that homophily explained over 50% of perceived assortative mixing/temporal clustering. The authors concluded that a lot of what was thought to be influenced-based contagion was in fact homophily-driven diffusion.

Online experiments and health behaviour

Centola (2010, 2011) uses online field experiments to study the social transmission of health behaviour in an attempt to uncover the exact processes by which people are influenced by others in their social network (Section 10.2.1 looks at this further, using observational data).

Centola (2010) asks whether some network structures are more or less effective in promoting social influence. The literature on this is at odds. On the one hand, the 'strength of weak ties' literature (Box 3.3) suggests that the diffusion of an innovation (e.g. information, behaviour) through a network will occur more efficiently where there is low redundancy of ties – that is, your neighbours do not tend to be connected to one another, and hence do not tend to share the same information. In this view, transmission of behaviour is a simple contagion process: if your neighbour exhibits the behaviour then you have a fixed probability of adopting the behaviour, and the probability does not increase with the number of your neighbours who have the behaviour.

On the other hand, Centola and Macy (2007) suggest that while the 'weak ties' argument might hold for innovations such as information or disease, it may not hold up for transmission of behaviours where social affirmation from multiple sources is required for successful transmission (what they term 'complex contagions'). Centola and Macy (2007) provide examples such as spread of high-risk social movements and avant-garde fashions. In such a view, it is likely that transmission will be aided by network structures that exhibit clustering, which increases the likelihood of contact from multiple sources.

Centola (2010) explored the role of network structure in the promotion of social influence by creating an Internet-based health community containing 1,528 participants recruited from online health communities of interest. Participants were randomly assigned health 'buddies' from whom they could gain information about a new online health forum. The act of joining the health forum was the health behaviour under examination, and the study aimed to quantify the impact of number of sources of information on the probability of adopting the behaviour. Participants were randomly assigned to networks with differing levels of clustering, which meant there was variation in the number of network neighbours providing information about the new health behaviour. The author found that the probability of adoption increased markedly when participants received social reinforcement from multiple network neighbours.

In the above discussion on social influence it was noted that homophily is a potential confounding effect in empirical attempts to identify social influence. That is, it is possible, using the words of Aral et al. (2009), to confuse homophily-driven diffusion with influence-driven contagion. However, it is possible that homophily, while playing a confounding role in the identification of social influence, also plays a direct role in social influence itself. That is, if two individuals are more like one another, then there may be a greater probability of transmission of behaviour than if they are different from one another. Thus, we need to look at 'homophily-driven contagion' or the role of dyadic attributes (e.g. homophily) as contributors to social influence (Section 10.2.1 looks at this further, but focusing on node-level and network-level attributes, rather than dyad-level).

Centola (2011) used the Internet-based health community described in Centola (2010) to conduct controlled experiments to study the effect of homophily on the adoption of health behaviour. The author argued that a highly homophilous social network may constrain adoptions of new behaviour because individuals tend to interact only with people like themselves, and hence are less likely to be exposed to innovations. Hence, if there is homophily on the basis of health status (unhealthy people tend to associate with other unhealthy people, while healthy people similarly seek out each other), this will mean that unhealthy individuals do not interact with healthier people, who may be the source of innovative health-related information. Countering this is the fact that homophily can increase dyadic-level social influence (people are more likely to adopt behaviour that they learn about from other 'similar' people).

The use of online experiments allowed Centola (2011) to identify the spread of a health innovation through fixed social networks that exhibited different levels of homophily. It was found that homophily led to a significant increase in overall adoption of a new health behaviour, especially among those most in need of the health innovation (the unhealthy).

Influence in Twitter

How has influence been measured in Twitter? Kwak et al. (2010) constructed three measures of Twitter influence (number of followers, PageRank and number of retweets), while Cha et al. (2010) used number of followers, number of retweets and number of mentions. In both papers, the authors found that there was no consistent ranking across the different measures for different users – a user with the highest influence in terms of PageRank rank did not necessarily rank highest on number of retweets.

In Twitter, how do we know when someone has been influenced, that is, what is the appropriate outcome? It appears that the best measure of influence in Twitter is retweeting, since this is a clear indication that someone has made a conscious decision to pass information on. Perhaps this is an appropriate measure of influence in an information network such as Twitter. As such, it would be possible to conduct a study similar to that of Katona et al. (2011) in the context of a social network site (Section 10.2.1), and use statistical techniques to identify the effect of ego network structure, characteristics of network neighbours who have already retweeted ('influencer effect'), and characteristics of the individual who is choosing whether to retweet ('adopter effect').

5.3 CONCLUSION

This chapter has tackled two big and related topics in social networks research: homophily and social influence. For both, the web is proving to be an important source of data for empirical research. With regard to homophily, we showed that while this is a relatively simple concept to grasp, measuring it accurately is challenging because of the existence of confounding factors and also limitations of data. We presented an example where Facebook has been used to study homophily, and the ability to collect more detailed information on social network participants led to new insights into this phenomenon. With regard to social influence, we also looked at the challenges with identifying whether a person's behaviour has been influenced by his or her social network, and we showed that social media data are being used in innovative ways to further our understanding about the nature of social influence.

Further reading

For more on social influence, see Christakis and Fowler (2009).

6

Organisational Collective Behaviour

Having focused in Chapter 5 on how people interact with one another on the web, we now look at organisational collective behaviour on the web. This chapter explores how web data are being used to understand organisational collective behaviour, with particular focus on organisations that are seeking to bring about social change.[1]

Section 6.1 provides some background to the use of the web for collective behaviour and the issues relating to the use of digital trace data for studying this behaviour. Section 6.2 examines a strand of research into online collective behaviour which comes out of political science research into public goods. Section 6.3 takes a more sociological perspective on collective behaviour, introducing the concept of networked social movements and how they can be studied using web data.

6.1 COLLECTIVE BEHAVIOUR ON THE WEB: BACKGROUND

Organisational collective behaviour refers to the pursuit of goals shared by a group of individuals, where the action is taken either directly or on behalf of the group through an organisation. We focus in this chapter on organisations that seek to enact social change. The term that is often used to describe such organisations is non-government organisations (NGOs), but equivalent terms are advocacy groups, civil society and grassroots organisations. In Section 6.3 we use the term that comes out of the social-movements literature – social movement organisation (SMO).

The web has become the major communication and organisation tool used by organisations seeking to enact social change. There are practical reasons for this: building and maintaining a web presence is relatively cheap, allowing organisations to remain visible even if 'there are few participants,

---

[1] In Section 9.1.1 the topic of collective behaviour is revisited, but the emphasis there is on collective behaviour at the individual rather than organisational level.

little activity is taking place or they are low on funds' (Pickerill, 2004, p. 181). The web also allows NGOs to bypass the mass media, which tend to highlight mainstream views (Castells, 1997; Tarrow, 2002 Garrett, 2006). Further, the web is a technology that can be used to enable values such as diversity, decentralisation, informality and grassroots democracy rather than centralisation and hierarchy (Pickerill, 2004; van de Donk et al., 2004), and so there are also important ideological and organisational reasons why NGOs have embraced this medium.

For the above reasons the web has become vital to collective behaviour, and Castells (2004) has asserted that social movements swim on the Internet 'like fish in water'. This has meant that these organisations are much more easily accessible for research purposes. Twenty years ago most researchers would have found it prohibitively expensive to study social movements outside of their own country, and would have needed to focus on a relatively small number of domestic organisations since the research typically involved travelling to interview the principals of the organisations. Today a researcher can conduct a web analysis of several hundred environmental activist organisations from all five continents without even leaving the comfort of his or her home town.

But this leads us to two important issues. First, it should be recognised that since online advocacy organisations are more accessible for research purposes, there is a risk that researchers may be tempted to overstate their importance (Rucht, 2004). Authors such as van de Donk et al. (2004), Bennett (2004) and Rucht (2004) have questioned the importance and impact of online social movements. For example, van de Donk et al. (2004, p. 18) suggest that online social movements may lack the attractions of group experience and the 'fun and adventure' factor that accompanies some forms of offline protest, thus lessening their appeal to potential participants. Since online social movements are easier to join and leave, and sometimes lack ideological coherence, it may be more difficult to organise coherent campaigns, as well as being more difficult to stop campaigns – a prime bargaining tool against campaign targets (Bennett, 2004).

We do not pursue this issue any further here. But given that a researcher has decided to use web data to study online collective behaviour, the second issue is how to do so. The focus of this chapter is on analysing online organisational collective behaviour using digital trace data – in particular, hyperlinks. As discussed in Chapter 2, analysing digital trace data in this way is an example of unobtrusive social research. In this mode of research, the researcher is not asking the organisations why they have chosen to place particular hyperlinks and text on their websites, but attempting to infer something about their organisational goals and behaviour by analysing the digital traces of these activities. An important assumption here is that the choices of the webmaster are accurate reflections of the organisational goals.

It was noted in Section 4.1.1 that there are many possible motives for an organisation to create a hyperlink to another organisational website. The emergence, operation and dynamics of groupings of actors who share a common goal of enacting social change have been the subject of much social scientific research. This chapter presents a brief review of two particular strands of research that have developed in political science and sociology, and shows how they have been adapted to the study of online collective behaviour. While the two theories of hyperlinking that have been advanced in the context of organisational collective behaviour (information public goods and online collective identity) have different disciplinary origins, they both can be empirically tested using ERGMs (Section 3.2.2).

6.2 COLLECTIVE ACTION AND PUBLIC GOODS

Much of the political science research into collective behaviour has involved the use of public choice theory (where concepts from economics are used to study political science phenomena). In particular, collective action has been modelled as a type of *public good*. Public goods exhibit two qualities: *non-rivalry* or *jointness of supply* (my consumption of the good does not affect your consumption) and *non-excludability* (no one can be effectively excluded from consuming the good).[2] For political scientists a key concern has been to understand the reasons why individuals and organisations contribute to collective action, rather than simply sitting back and benefiting from the efforts of others – the 'free rider' problem (Olson, 1965).

6.2.1 Hyperlink networks as information public goods

Fulk et al. (1996) and Bimber et al. (2005) extend the definition of public goods to include information and computer-mediated goods. *Information public goods* are therefore defined as including, for example, publicly available electronic databases, and Section 9.1.1 provides other examples of information structures on the web that can be thought of as information public goods.

NGO hyperlink networks and the North-South divide

Shumate and Dewitt (2008) argue that a hyperlink network formed by NGOs working on a particular issue is an example of an information public good. Hyperlinks enable actors (whether they be NGOs working on the same or other issues, government agencies, corporations or members of the public) to locate and make sense of the purpose and position of the NGOs focused on this particular issue, either by following hyperlinks within the network or else via search engines such as Google. An NGO hyperlink network therefore exhibits jointness of supply since the act of one person searching the network (either using a search engine or by following hyperlinks) does not preclude

---

[2]Classic examples of public goods are national defence, free-to-air television and radio, roads and street lights.

others from doing the same. Since the hyperlink network is on the public web it is also characterised by impossibility of exclusion – all users with a computer and Internet connection can access the hyperlink network.

Shumate and Dewitt (2008) used a snowball sampling approach (Section 2.2.2) to collect data on the organisational hyperlink network formed by 248 NGOs focused on HIV/AIDS. About 75% of the NGOs were from the geopolitical 'North' (North America, Western Europe, Australia and New Zealand) while the remainder were from the 'South'. Previous authors have argued that the web has transformed spatial relations among NGOs, leading to a transcendence of the North–South divide, and Shumate and Dewitt (2008) test this hypothesis by statistically analysing the hyperlink network formed by the HIV/AIDS NGOs.

The authors proposed several hypotheses – which they tested using ERGMs (one of the first applications of ERGMs to hyperlink networks). First they hypothesised that the hyperlink network would exhibit a significant level of reciprocity; this hypothesis was supported in the ERGMs. Reciprocity in the NGO hyperlink network accords with their theory that NGO hyperlink networks are examples of information public goods: NGOs are hyperlinking to one another to improve the 'searchability' of the network, thus creating an information public good.

They then tested a series of hypotheses that relate to the North–South configuration of the hyperlink network. They argued that Southern NGOs are more likely to concentrate on local and grassroots efforts and thus are less likely to seek coalitions that might be reflected in the hyperlink network. In contrast, Northern NGOs are more likely to focus on national and international partnership making. Overall, contributions to the hyperlink network are more likely to be made by Northern NGOs.

These findings are relevant to globalisation research (Box 6.1). Shumate and Dewitt hypothesised that if it is the case that Internet technology is transforming spatial relations, then an NGO's position in a hyperlink network will not reflect its geographic location. They found that (contrary to the expectations that follow from research into globalisation and the Internet), hyperlink networks are shaped by the North–South divide. The majority of NGOs in their sample were from the global North, and these NGOs are statistically more likely to form ties with other NGOs from the global North. Drawing on Rogers and Marres (2000), they argued that NGOs from the North are drawn to the centre of the hyperlink network.

6.3 NETWORKED SOCIAL MOVEMENTS

Sociologists have advanced two main approaches for understanding collective behaviour of organisations seeking social change.[3] The *resource mobilisation*

---

[3]This section draws from Ackland and O'Neil (2011).

BOX 6.1   INTERNATIONAL HYPERLINK NETWORKS

World-systems theory (e.g. Wallerstein, 1974) emphasises the inter-dependence of nations on the basis of flows of goods, labour, capital and information, and this interdependence can be formalised as a network. The theory posits that countries occupy one of three positions in the network: core, periphery and semi-periphery (e.g. Park et al., 2011), with particular countries (e.g. the US) occupying dominant positions (hubs) in a highly centralised network, with other countries (e.g. those in the global South) as spokes connected to the hub but not to one another.

However, it is an empirical question whether the network of countries is highly centralised, with a core, periphery and semi-periphery structure. While world-systems theory emphasises the role of economics in structuring relations between countries, other forces such as culture, language and geography could counteract economics, leading to the emergence of a more decentralised network with clustering in peripheral areas (e.g. Barnett, 2001; Lee et al., 2007). The testing of world-systems theory thus involves using network science to establish the properties of networks of countries (How centralised is the network? Does the network have the core/periphery/ semi-periphery structure that the theory predicts?) and the position of individual countries within the network (what nodes are at the core and the periphery?). There is also a need for longitudinal analysis to establish whether the level of centralisation is changing over time and whether there are significant changes in the positions of particular countries.

World-systems theory has been tested using international telecommu-nications traffic (e.g. Barnett, 2001) and, more recently, hyperlink data (e.g. Park et al., 2011). An international hyperlink network is a network of countries where weighted ties between countries reflect the sum of hyperlinks directed between websites based in each country. There are methodological challenges involved in constructing international hyper-link networks, such as the fact that in some countries there is a tendency not to use the country code TLD, and it is consequently difficult to ascribe hyperlinks pointing to or from such websites (e.g. .com websites) as relating to a particular country. There has been some recent progress in 'cracking the .com domain', resulting in significant changes to the structure of inter-national hyperlink networks (Barnett et al., 2011).

approach models individuals and groupings of individuals as rational actors engaged in strategic or instrumental behaviour, such as forming alliances with other actors in pursuit of their interests (McCarthy and Zald, 2002). In contrast, *new social movement theory* emphasises expressive (rather than instrumental) behaviour leading to a shared sense of identity between social actors, without which there can be no collective behaviour to enact change (Melucci, 1995).

Mario Diani has made important contributions to social movement theory and the use of social network analysis to study social movements. Diani (1992) defined a social movement as a grouping of actors who are engaged in conflict or competition over a social problem and who have two additional important characteristics.

First, these actors share a *collective identity*, which is a mutually agreed-upon (and often implicit) definition of membership, boundaries, activities and norms of behaviour used to characterise a grouping of actors. According to Snow (2001, p. 2213): 'Discussions of [collective identity] invariably suggest that its essence resides in a shared sense of "one-ness" or "we-ness" anchored in real or imagined shared attributes and experiences among those who comprise the collectivity and in relation or contrast to one or more actual or imagined sets of "others".' Social-movement researchers have emphasised the importance of *frames* for the development of collective identity; Evans (1997, p. 454) argues that 'collective identities result from framing processes'. A frame is a 'schema of interpretation' (Goffman, 1974, p. 21) which allows for a social problem to be identified and addressed, thus guiding individual and collective action (Benford et al., 1986). An example is the 'Frankenfoods' frame that was used in protests against genetically modified organisms.

Second, Diani (1992) argues that social movement actors exchange practical and symbolic resources through informal networks. Practical resources can be valued or measured objectively (e.g. money, members). Hoffman and Bertels (2007) construct a network of NGOs using the 'interlocking directorate' approach (where a tie between two organisations reflects the fact that they have one or more people in common on their board of directors), and this network can be seen to enable the transfer of practical resources such as information and funding. The exchange of symbolic resources is a means of establishing boundaries of inclusion/exclusion and hence is related to the formation of collective identity. Diani and Bison (2004, p. 298) constructed a network showing the exchange of symbolic resources between social movement organisations by assessing whether the voluntary organisations in their study 'feel links to their partners ... [which] imply some kind of broader and long-term mutual commitment? Do they, in other words, share a collective identity?'

Online collective identity

Ackland and O'Neil (2011) present a conceptualisation of an online social movement that aims to extend the Diani network-theoretic approach to the online world. The authors define an online social movement as a set of websites of organisations which are engaged in competition over a social problem and which: (1) share a collective identity; (2) exchange practical and symbolic resources via the use of hyperlinks and website text content.

Ackland and O'Neil (2011) contend that two types of resource are exchanged in SMO hyperlink networks. First, 'index authority' is a practical resource, which is what a website gets when other relevant sites link to it. As discussed in Section 7.1.1, inbound links from relevant sites translate to higher ranking in search engine indexes (such as Google) and hence greater online visibility. Second, hyperlinks also facilitate the transfer of a symbolic resource that helps establish 'boundaries of belonging' and hence collective identity.

BOX 6.2   OFFLINE CHARACTERISTICS OF NGOS AND HYPERLINK NETWORKS

In Ackland and O'Neil's (2011) analysis of hyperlinks and online collective identity, the focus was on SMO attributes that relate to collective identity, with websites being categorised according to the main issues or causes. While some geographical information was also used in the analysis (websites were categorised according to whether the organisation is located in the geopolitical North or South), the primary focus was on information about sub-movement identity information that could be garnered directly from the web, rather than offline characteristics.

However, other authors have studied how *offline* characteristics structure NGO hyperlink networks. For example, Gonzalez-Bailon (2009) used ERGMs (Section 3.2.2) to model the hyperlink network of 967 NGO websites and found that offline economic resources (proxied by number of paid staff) and visibility (proxied by LexisNexis archive data on total number of mentions in major world publications, newspapers, TV and radio) had a significant effect on the probability of receiving hyperlinks. (Using the ERGM terminology introduced in Section 3.2.2, there are significant receiver effects for offline economic resources and visibility.) The results led to a conclusion that 'some agents enter the web from a position of strength that does not derive from their online activities but from their access to economic resources and offline visibility' (Gonzalez-Bailon, 2009, p. 279). This is relevant to discussions in Sections 7.1.1 and 9.2.2 on the 'democratising' potential of the web.

The authors contend that to the extent that the belief systems of social movements have been institutionalised in the online environment, we should see evidence of this in the digital trace data created by their activities.[4] The hyperlink network should therefore exhibit particular structural signatures of online collective identity. In particular, they propose that a given set of websites run by organisations engaged in competition over a social problem can be regarded as an online social movement if the hyperlink network exhibits significant informal, endogenous or purely structural network effects, and if there is significant homophily on the basis of sub-movement affiliations.

The authors identified 161 environmental activist websites using a combination of snowball sampling and search techniques proposed for researching issue networks (Section 4.2.2; Rogers and Zelman, 2002). The VOSON software (Section 4.3.1) was used to automatically collect the hyperlinks between the sites and also text content (meta keyword) data from the homepages. The

---

[4] Hunt and Benford (2004, p.414) note there is a tendency for social movement scholars studying collective identity to 'appear to take for granted [its] existence without offering compelling evidence that [it exists] outside the minds of the social movement analysts'.

sites were manually coded on the basis of three hypothesised sub-movements: 'globals' are organisations that are focused on a range of issues such as climate change, forest/wildlife preservation, nuclear weapons and sustainable trade; 'toxics' are focused on pollution and environmental justice; and 'bios' are mainly concerned with genetic engineering, organic farming and patenting issues. See Box 6.2 for examples of research into NGO hyperlink networks where offline characteristics of the NGOs are included in the analysis.

The ERGM results showed there was significant homophily over the hypothesised sub-movement classification, allowing the authors to conclude that this collection of environmental activist websites comprised an online social movement. They also noted the existence of an apparent structural hole (Box 3.4) between bios and toxics in the hyperlink network, which they saw as evidence that class distinctions may be playing a role in structuring the online collective identities of activist networks.

6.4 CONCLUSION

There is a tradition of researching offline organisational networks (e.g. inter-locking board directorates) using social network analysis, and this chapter has explored how SNA can be used to study NGO hyperlink networks. The chapter started by looking at why so much organisational collective activity has shifted onto the web, and how this has presented both opportunities and challenges for researchers. The next two sections presented two conceptual frameworks for studying NGO hyperlink networks – NGO hyperlink networks as information public goods and NGO hyperlink networks as representations of online collective identity. While these two approaches here have different disciplinary origins, they are both amenable to statistical network analysis (in particular, ERGMs) and hence they can be seen as examples of social hyperlink networks from Chapter 4.

Further reading

For a theoretical approach to networked social movements, see Castells (2004). For more on the study of social movements using network techniques, see Diani and McAdam (2003) and della Porta and Diani (2006). Smith and Kollock's work (1999) is an early example where collective action theory is adapted to online settings. O'Neil's (2009) perspective on spontaneous organisation of groups online is also relevant.

7

Politics and Participation

A lot of social science research into the web has been motivated by the need to understand whether online behaviour can have influence in the real world, that is, whether there is a social impact of the web. This chapter looks at the evidence of how the Internet may be affecting the visibility of political information (and hence the balance of political power between minor and major political actors) and individual-level engagement with politics. The chapter also considers how Internet use may be changing patterns of social connectivity.

Section 7.1 looks at how hyperlink data can be used to evaluate whether the web has created a level playing field in politics, with minor and fringe political actors competing on equal terms with the major parties. The section introduces the topic of power laws, which can be used to describe the distribution of attention on the web. Section 7.2 shifts the focus to the individual, providing an assessment of research into how Internet use impacts on social connectivity (studied via the concept of social capital) and political participation. Section 7.3 looks at empirical evidence on whether the web has led to a shift in the extent to which people are likely to 'engage with the other side' when it comes to politics, or instead only connect with individuals who share similar political beliefs (thus clustering on the basis of political identities or attitudes).

7.1 VISIBILITY OF POLITICAL INFORMATION

An early Utopian view of the political impact of the web was that it would create a level playing field, enabling minor and fringe political actors to compete on equal terms with the major parties. This section looks at how hyperlink data can be used to assess whether this has happened.

7.1.1 Power laws and politics online

It was predicted that the web would enable those with non-mainstream political views to have a 'voice'; early predictions were that the web would 'ameliorate inequalities of attention to views and information sources that are outside of the political mainstream' (Hindman et al., 2003, p. 3). This is related to arguments about the 'democratising potential' of the web in other contexts, for example, whether the web is leading to a 'reconfiguring' of access to scientific authority (Section 9.2.2).

How can we empirically assess whether the web has altered the distribution of attention away from mainstream political viewpoints and messages? Even if survey data measuring public perceptions or knowledge of alternative political messages were available over time, it would be very difficult to establish that a change in public awareness of political information was caused by the advent of the web, rather than some other factor. Arguably, the most accurate measure of attention being directed to a website is the number of visitors to the website over a given period of time – otherwise known as hits, clicks, eyeballs or visits. However, it is difficult to accurately collect such data since this typically requires access to the website log files, which are not publicly available. So instead, researchers have focused on an indirect measure of the distribution of political attention: hyperlinks to websites containing political information.

Does hyperlink data have construct validity (Section 1.5) in this context? That is, can we argue (or, even better, show empirically) that measures constructed from hyperlink data are likely to be strongly correlated with our construct – attention to political information and viewpoints? If we accept that website hits have construct validity, then we next need to show that hyperlinks are correlated with website hits.

Hindman et al. (2003) provide compelling arguments for the construct validity of hyperlinks in the context of the visibility of political information. First, they argue that people generally find new online information either by following hyperlinks from pages they are currently reading or else via search engines such as Google. Second, the probability of a given page being found via either of these techniques is dependent on the structure of web graph (a network where web pages are nodes, and hyperlinks are ties) – in particular, the number of inlinks to that page from other relevant pages. While web page *retrievability* is absolute and equal across all pages (all web pages are equally retrievable, assuming the web server is running), web page *visibility* is a relative concept and is largely determined by the link structure of the web.

If we accept that hyperlinks are viable indicators of political attention, how can we use them to test whether the web has led to a redistribution of political attention from mainstream political actors to the fringes of the political debate? Two approaches have been proposed. First, research has looked at counts of inbound hyperlinks to major and minor political party websites, in the context of the *normalisation thesis* (Box 7.1). This is an application of webometrics (Section 4.2.1) to political science research. The second approach employs techniques from applied physics, and involves the study of power laws on the web.

Political 'Googlearchy'

Hindman et al. (2003) focused on the question of whether the web has led to a reduction in the cost of acquiring political information. This is a different question than that posed above (i.e. whether the web has changed the distribution of political attention), albeit a related one. Whether the web has reduced the cost of acquiring political information is in fact an intermediary question. That is, Hindman et al. (2003) argued that if the relative cost of gaining political information from non-mainstream sources is decreased by the advent of the web, then one would expect that the distribution of political attention to mainstream and non-mainstream political actors would become more even.

Hindman et al. (2003) proposed that the question of whether the web has led to a more equal distribution of political attention can be answered by looking at the distribution of hyperlink indegree for politically oriented web pages. Applied physicists (e.g. Barabási and Albert, 1999; Barabási et al., 2000) have concluded that the distribution of indegree in many large-scale networks, including the web, follows a *power law*. The existence of a power law means that the counts of hyperlinks to sites follow a very unequal distribution, with a very small number of sites receiving the lion's share of links, and the majority of sites receiving very few or no links. See Box 7.2 for an explanation of why power laws emerge.

BOX 7.2   HOW POWER LAWS DEVELOP (PREFERENTIAL ATTACHMENT)

Having discovered that power laws are one of the *empirical regularities* of large-scale networks such as the web, applied physicists then tackled the question of how such power laws emerge. Barabási and Albert (1999) developed the model of *preferential attachment*. In a growing network, new entrants to the network have a preference for creating links with existing popular nodes. This leads to a 'rich-get-richer' phenomenon, exemplified by power laws in link distributions, with older nodes having a greater number of inbound hyperlinks.

While the preferential attachment model has been influential, it leads to one conclusion that is apparently at odds with empirical evidence: the model predicts a positive correlation between the age of a site and its capacity to attract links, while Adamic and Huberman (2000) studied a crawl of 260,000 sites and found no such correlation between age of site and indegree. Bianconi and Barabási (2001a, b) extended the basic preferential attachment model by incorporating the concept of 'fitness' whereby the probability of acquiring new edges is also a function of a node's intrinsic worth, and not just its current degree. This modification results in node age and degree no longer being correlated.

Pennock et al. (2002) measured the distribution of inbound hyperlinks and traffic and found that while the connectivity distribution of the entire web may resemble a power law, this was not necessarily the case for sub-categories of web pages (e.g. university and newspaper home pages). Hindman et al. (2003) used an innovative text mining approach to identify politically oriented web pages (on topics such as abortion, the death penalty, gun control), and showed that amongst these subsets of web pages the distribution of inbound hyperlinks *does* follow a power law, suggesting that these information environments are dominated by a few websites (leading to political 'Googlearchy'). This suggests that the cost of acquiring political information may only have been reduced for those wishing to acquire information from mainstream political actors, and hence the web may not have led to a democratising of access to political information.

But Hindman et al. (2003) noted that the *narrowcasting* nature of the web means that it is possible for a person to easily and regularly acquire political information from non–mainstream sources on the web even if those sources (e.g. fringe political groups) are in the Long Tail (Section 10.1), that is, not heavily linked to and therefore not returned highly in Google searches. Also, the fact that search engines operate on the basis of text queries means that a person looking for information on a non-mainstream political topic might be able to easily locate relevant websites in Google searches simply because mainstream websites do not mention the text that is being searched for (this relates to the discussion about the fragmentation of scientific fields in Section 9.2.2).

7.2 SOCIAL AND POLITICAL ENGAGEMENT

In this section we look at research into the web and social and political engagement.

7.2.1 Web use and social capital

Social capital is an influential concept that originated in sociology (Box 7.3). There is a large body of research into the relationship between Internet use and social capital. A motivating factor for this research was the suggestion (e.g. Putnam, 2000) that younger people have less social capital (measured by community volunteerism and trust in fellow citizens) than their parents did when they were young. Since younger people use the Internet more than older people, this had led to the question of whether Internet use decreases social capital (the 'moral panic' about the social impact of the Internet has been particularly evident in the area of online games – Box 7.4). McPherson et al. (2006) found that there had been a decline in the size of 'core networks' in the US (people with whom the respondent can talk about 'important matters') and an increase in social isolation.[1] These findings reignited panic about the negative effect of the Internet on social connectivity, even though the authors and others (e.g. Fischer, 2009; McPherson et al., 2009) noted that the results indicating a trend of increasing social isolation needed to be treated with caution.

So what is the evidence on the impact of Internet use on social capital? Early research found there was a negative relationship (Kraut et al., 1998), helping to fuel the moral panic, but a follow-up study (Kraut et al., 2002) did not support this earlier finding. Shah et al. (2001) is an early example of research into the impact of Internet use on social capital where a distinction was made between various types of Internet use (previous authors had tended to group all types of Internet activity into a single variable, hours of use). The authors measure an individual's social capital using two indicators: engagement in community activities and trust in others (an example of

---

[1]A person's core network is generally composed of 'strong ties' (people with whom the person shares friends in common, or family members) and is an important source of social support (Wellman and Wortley, 1990), which can assist in overcoming health problems and other adverse events (e.g. Cohen, 2004).

BOX 7.3 DEFINITIONS OF SOCIAL CAPITAL

There are two main ways that social capital has been defined. The first approach, which is mainly identified with the work of Nan Lin (e.g. Lin, 1999, 2001), attempts to operationalise the value of an individual's social network: 'Social capital can be seen as similar to human capital in that it is assumed that such investments can be made by individuals with expected return ... to the individual' (Lin, 1999, p. 32). Other relational-level definitions of social capital are: 'the number of people who can be expected to provide support and the resources those people have at their disposal' (Boxman et al., 1991, p. 52) and 'the ability of actors to secure benefits by virtue of membership in social networks or other social structures' (Portes, 1998, p. 6).

The second approach to defining social capital operates more at the group or community level, and is primarily identified with Robert Putnam, who defines it as 'features of social organization such as networks, norms, and social trust that facilitate coordination and cooperation for mutual benefit' (Putnam, 1993, p. 35). A related definition is that of Fukuyama (1999a, p. 16): 'Social capital can be defined simply as the existence of a certain set of informal values or norms shared among members of a group that permit cooperation among them.'

The relational-level concept of social capital is more easily quantifiable at the level of the individual, compared with the group-level definition, and involves identifying 'resources embedded in social structure which are accessed and/or mobilized in purposive actions' (Lin, 1999, p. 35). Survey-based techniques for constructing individual-level measures of social capital involve the name generator, position generator and resource generator (e.g. see, van der Gaag and Snijders, 2005).

While the concept of social capital has been extremely influential, there are some acknowledged problems with its measurement. There are many competing ways of measuring social capital, and, as Fukuyama (1999b, p. 12) states, 'one of the greatest weaknesses of the social capital concept is the absence of consensus on how to measure it', although this is possibly more of a problem with the group-level rather than relational-level definition of social capital. A major problem with research that aims to demonstrate the impact of social capital is the identification of causality: is a good outcome for an individual caused by the person's social capital or is there merely correlation with other unobserved attributes (selection effects)? See the related discussion in Sections 5.2.1 and 9.3.1.

operationalising the group-level definition of social capital, rather than the relational-level definition – Box 7.3).

Internet use was classified into four types: social recreation (playing online games, participating in chat rooms or online forums), product consumption (e.g. purchasing a book or video), financial management (online banking), and information exchange (sending email or searching for information on

BOX 7.4 WHO IS THE 'AVERAGE' ONLINE GAMER?

There have been small-scale ethnographic studies of player guilds and communities in MMORPGs, but few large-scale and generalisable quantitative analyses. As a result, the profile of the average online gamer (demographics, mental and physical health) is subject to stereotypes and moral panic. The stereotype is that gamers are young males with no social skills and few offline social connections.

Williams et al. (2008) undertook the first large-scale analysis of online gamers. The authors collected data from EverQuest 2 (EQ2) and matched a socio-demographic survey (obtrusive data collection) with data on online behaviour (unobtrusively collected game logs). The demographic survey was a random sample, rather than a convenience sample. While the survey was of EQ2 players, the researchers had to reach the players via their characters (avatars) and hence the sampling frame was populated with characters. Players sometimes have more than one character (e.g. on different EQ2 servers), but research shows that such players typically have a main character, and this can be identified from the server logs. So each player had the same chance of being sampled, regardless of the number of characters they may play.

Players with characters in the sampling frame were invited to participate in a 25-minute survey. Sony created a special game object that was given to participants (the 'Greatstaff of the Sun Serpent'), and the target of 7,000 respondents was achieved in two days. The results showed that the mean time spent playing per week was 25.9 hours, and the mean player age was 31.2 years, which was higher than expected. Confirming one of the stereotypes of online games, 80.8% are male, but females play longer (29.3 hours per week compared with 25 hours for men). Whites and Native Americans are more likely to play, while Asians, blacks, Hispanics/Latinos are less likely, but there was no difference in hours of play, by race.

Compared with US Census data, EQ2 players come from wealthier backgrounds. They are less spiritual in general, and less likely to be from a mainstream religion when they do practise. Physically, EQ2 players are healthier than average, but players have lower levels of mental health on two out of three of the mental health indicators used in the survey. But the authors noted the difficulty of ascribing causality in these results.

the web). The authors found that overall Internet use and social recreation Internet use are not significantly correlated with social capital, while information exchange is associated with a significant increase in social capital. The analyses were conducted separately for different demographic groups (Generation X, baby boomers, civic generation). For Generation X, there was a negative correlation between social recreation use of the Internet and social capital measured as interpersonal trust (but there was no significant relationship with community volunteerism).

The authors acknowledged that causality cannot be inferred from these results. The difficulty of establishing causality is a common limitation of research into Internet use and social capital – we look at this later in the context of political engagement (Section 7.2.2). People who are heavily involved in their communities may use the Internet for information exchange in order to fulfil their pre-existing duties to community organisations. Similarly, among those in the Generation X group in the sample, it is plausible that someone who is innately mistrustful of other people might start using the Internet for social recreation purposes because they do not want to have face-to-face social interactions.

Zhao (2006) used the number of 'voluntary' social ties (not ties with family or work colleagues) as an indicator of social capital (an example of the relational-level definition of social capital). The definition of social ties included both offline and online ties; as early as 1999, Nan Lin argued that it is necessary to take account of online social ties in measurement of social capital, predicting that 'we are witnessing a revolutionary rise of social capital, as represented by cyber-networks' (Lin, 1999, p. 45). Zhao (2006) also distinguished between non-social use of the Internet (web surfing) and social use (e.g. email, chat rooms), and confirmed previous research findings that non-social Internet users do not have fewer social ties (compared with non-users). It was also found that social Internet users have more social ties than non-users, and that heavy chat users have more social ties than light chat users (the same is found for email users). However, the positive correlation between social Internet use and social capital could be an artefact of the data, since the measure of social capital includes online social ties.

While much of the research into the Internet and social capital has focused on how Internet use changes the *size* of personal and core networks, the impact on a person's *diversity* of social connections has also been a focus of research. This research is often couched in terms of 'bridging' and 'bonding' social capital (Putnam, 2000), where the former refers to the new information and resources that a person can gain from having (often weaker) connections to diverse social groups, and the latter refers to the emotional and substantive support that can be accessed in the person's core network (predominantly consisting of strong ties). The distinction between bridging and bonding social capital is thus clearly related to the 'strength of weak ties' concept (Box 3.3) and structural holes (Box 3.4).

The Internet can clearly aid the formation of bridging social capital since it is possible to overcome geographic and other constraints and connect with people in virtual spaces. The study by Williams et al. (2006) of World of Warcraft guilds provided evidence of bridging social capital formation. One interviewee remarked, 'I've become closer to some of my [real-life] friends thanks to WoW ...' cause it gives u more to talk about and shared experiences' (p. 351), while another said, 'I suppose in a way this is a way for us to socialize and do things together despite distance. I think

WoW just makes it easier for us to keep in touch as old friends and do something together' (p. 351).

While there have been suggestions that the Internet may foster the creation of bridging social capital at the expense of bonding social capital, the evidence does not support this (at least in terms of the influence of Internet use on core network size). In a recent study, Hampton et al. (2011) supported the conclusion of McPherson et al. (2006) that there has been a decline in the size of core networks in the US, but found no evidence that Internet users have smaller core networks compared with non-users. Williams et al. (2006) also provided evidence of bridging social capital formation in WoW, with one interviewee commenting: 'I have more in common with my online friends than I do my offline. RL friends are limited to who I've met at work or thru other work friends. On the internet, I have a MUCH larger pool of people' (p. 352).

7.2.2 Political engagement

Political engagement or *political participation* refers to activities that are aimed at contributing to a political outcome, and include: voting (and persuading others to vote); displaying buttons, signs or stickers; making campaign contributions; volunteering for candidates or political organisations; contacting officials; contacting media; protesting; participating in petitions (email or written); and boycotting. Political scientists have argued that the Internet increases political engagement by reducing both the direct and opportunity cost of acquiring political information (e.g. DiMaggio et al., 2001; Norris, 2001).

Research into factors influencing political participation has emphasised the role of socio-economic characteristics (education and income), demographic characteristics (age, sex, ethnicity) and also attitudinal factors such as partisanship and political interest (see Tolbert and McNeal, 2003, for a summary). These studies have generally found that mass media such as television and radio do not affect political participation, and Bimber (2001) found that this was also the case for Internet use. However, Tolbert and McNeal (2003) argued that the finding of Bimber (2001) might be due to the fact that different types of Internet use were not distinguished, and they provide evidence that using the Internet for accessing news has a significant positive impact on political participation, even after controlling for the other factors.

A meta-analysis by Boulianne (2009) of 38 US studies found that a majority show a positive correlation between Internet access and political engagement, but the author emphasised that the issue of causality is yet to be resolved. That is, is Internet access causing an increase in political engagement or else is the Internet simply replacing traditional political communication, with no influence on the structure and level of political engagement in the overall population (Bimber, 2001; Norris, 2001, 2005; Best and Krueger, 2005)? If it is the latter, then the positive correlation between Internet use and political engagement is arising from *unobserved heterogeneity*; that is, people who are already politically motivated are making use of the Internet for political engagement, and hence there is spurious correlation between

Internet use and political engagement.[2] This problem of distinguishing causation from correlation is endemic in empirical social science research, and is discussed in Sections 5.2.1 and 9.3.1. Box 7.5 presents an example of research where panel data and instrumental variables are used in an attempt to establish a causal relationship between Internet use and political engagement.

BOX 7.5 INTERNET USE AND POLITICAL ENGAGEMENT: ESTIMATION APPROACHES

Establishing whether Internet use causes increased political engagement is difficult with cross-sectional data. However, the use of panel (longitudinal) data allows the researcher to directly observe how political engagement of particular individuals changes with increased Internet use. Since the *same* individual is being observed at two different time points, it is more likely that any correlation between Internet use and engagement is in fact causation, since we can assume that a person's latent level of political interest/motivation does not change over time.

Kroh and Neiss (2009) use the German Socio-Economic Panel Study (SOEP) to study how Internet use affects four measures of political engagement: affiliation with a party, general interest in politics, active work in politics (e.g. as volunteer) and voting in elections. They use the 1995 and 2008 waves of the panel to estimate a fixed-effects model, which is used when the unobserved heterogeneity (in this case, latent political interest/motivation) is time-invariant and correlated with the independent variable of interest (in this case, Internet use) and hence can be effectively removed from the model via differencing. The use of longitudinal analysis led the authors to conclude that 'most of the cross-sectional correlation between political engagement and Internet access is attributable to unspecified background variables and self-selection processes of politically active citizens into Internet use' (p. 15).

Kroh and Neiss (2009) were also able to make use of regional variation in the introduction of broadband in Germany to conduct an instrumental-variables regression using the 2005 cross-section. They argued that since in 2005 some citizens were unable to access broadband simply because of where they live; this meant that 'access to broadband' could be used as an instrument for the Internet-access endogenous, independent variable. In instrumental-variables regression an instrument is a variable that is uncorrelated with the unobserved heterogeneity (in this case, latent political interest/motivation) but correlated with the endogenous independent variable (in this case, Internet use). While the authors provide caveats about the suitability of broadband availability as an instrumental variable, their results indicate that Internet use may in fact have a *negative* influence on political engagement.

---

[2] This spurious correlation could also occur because of omitted-variables bias – that is, if the model does not include variables that are likely to be correlated with both Internet use and political engagement, for example income and education, then there could be an incorrect conclusion that Internet use is causing increased engagement.

7.3 POLITICAL HOMOPHILY

Around the turn of the century, two radically different predictions regarding the web and politics were advanced. For some (e.g. Castells, 1996), the web would lead to a new era of *participatory democracy* (broad participation in the direction and operation of the political system). In contrast, authors such as Putnam (2000) and Sunstein (2001) argued that the web would lead to increased isolation and the loss of a common political discourse, leading to *cyberbalkanisation* (Van Alstyne and Brynjolfsson, 2005) – a fragmenting of the online population into narrowly focused groups of individuals who share similar opinions and are only exposed to information that confirms their previously held opinions.

It was noted above that the relative ease of joining virtual communities should help foster the formation of bridging social capital, which should work against cyberbalkanisation. However, virtual communities are generally organised around particular interests, and so, while it may be that the Internet can foster the collection of more weak ties (connections with people with whom you do not share friends), this does not necessarily increase bridging social capital (and at an aggregate level, reduce fragmentation). What is important is the nature of the virtual community. Joining a virtual community for cat lovers might increase your weak ties and the potential diversity of your social network but have no meaningful influence on your bridging social capital since it is unlikely that online conversations will be on topics other than cats. A person's bridging social capital may also be diminished via membership of virtual communities focused on political or social issues: while the person may be accumulating more weak ties, these are with other people who share similar opinions on important dimensions, and hence this would not increase bridging social capital.

The fragmentation hypothesis has been empirically evaluated using survey data. For example, Williams (2007) evaluated whether Internet use leads to an atomisation of social groups (less bridging social capital) and increased hostility towards 'out-groups', and found that the opposite was the case. While core networks are primarily seen as a source of bonding social capital, Hampton et al. (2011) note that core network members are influential in opinion and attitude formation, and hence core network diversity is important for exposing people to different ideas and opinions. The authors found that the use of particular types of social media (in particular, photo-sharing applications) was associated with an increase in the likelihood of having a person with a different political affiliation in the core network, but the authors noted that this might be due to 'pervasive awareness': finding out new things about friends via social-media status updates. The authors concluded that 'New technologies may not increase diversity as much as they increase awareness of diversity that was always there' (p. 150).

The fragmentation hypothesis can also be tested using digital trace data.

7.3.1 Divided they blog

While it seems unlikely that either of the opposing predictions of the future of political affiliation in the Internet age are likely to eventuate (it will probably be something in between), it is useful to consider how web data can be used to identify a movement towards or away from cyberbalkanisation.

Adamic and Glance (2005) analyse linking behaviour in the political blogosphere by constructing two datasets. The first was a single day's snapshot of around 1,500 political blogs (750 liberal and 726 conservative) collected on 8 February 2005.[3] In constructing this dataset, Adamic and Glance (2005) extracted all hyperlinks from the front page of each blog – there was no distinction between hyperlinks made in blogrolls (blogroll links) and those made in posts (post citations). The second dataset contains posts of 40 influential or 'A-list' bloggers (20 conservative and 20 liberal) collected during the two months preceding the 2004 US Presidential election. Adamic and Glance (2005) collected 12,470 posts from the liberal bloggers and 10,414 posts from the conservative bloggers that were made during the period 29 August to 15 November 2004, and they used text mining algorithms to separate out the post citations from the blogroll links. Their dataset only included the post citations (the authors argued that these are more indicative of blogger 'reading activity' than blogroll links, which can grow stale).

The authors found that while there was some cross-linking between conservatives and liberals, political bloggers predominantly hyperlinked to bloggers of the same political persuasion, leading to political homophily (Section 5.1) – the 'divided they blog' phenomenon that Adamic and Glance (2005) strikingly represented in a visualisation of the blogger network. Homophily on the basis of political or social issues has also been studied in other contexts. For example, Ackland and Evans (2005) constructed the hyperlink network of around 200 websites that were actively participating in the abortion debate in Australia and found strong evidence of pro-choice and pro-life clusters or communities (Figure 7.1).[4]

Another focus of the research into political hyperlinking has been whether conservative and liberal political actors might display different hyperlinking behaviour. Adamic and Glance (2005, p.1) found 'differences

---

[3] The one-day snapshot dataset is publicly available at http://www-personal.umich.edu/~mejn/netdata/polblogs.zip

[4] This is an example of the use of force-directed graphing to visualise a hyperlink network. Websites are initially randomly placed and then they act as if they are electrostatically charged (repulsion forces pushing them apart from one another). Hyperlinks between websites are modelled as springs (attraction forces pull together those sites that are hyperlinked to one another). The algorithm minimises the energy of the system by shifting the position of nodes, and this can lead to the emergence of clusters of websites that link more densely to one another than to sites outside of the cluster.

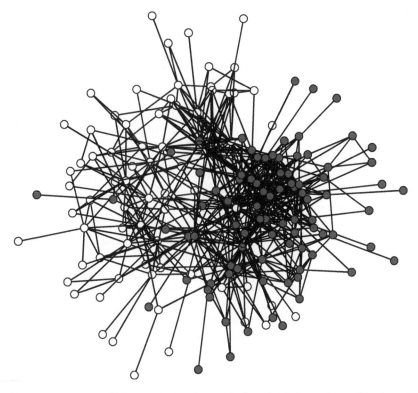

Figure 7.1 Hyperlink network of pro-choice (white) and pro-life (grey) websites (Ackland and Evans, 2005)

in the behavior of liberal and conservative blogs, with conservative blogs linking to each other more frequently and in a denser pattern'. In a study of the web presence of US participants in the abortion debate, Adamic (1999) also found that the network of pro-life websites is more 'tightly-knit' compared with the pro-choice network. Ackland and Shorish (2009) used the data from Adamic and Glance (2005) to further explore political homophily in the blogosphere, focusing in particular on whether conservatives and liberals exhibit different levels of homophily. They also developed an economic model of link formation that can be used to explain political homophily by relating the link formation behaviour of bloggers to the underlying distribution of political preferences.

The political behavioural study of Alford et al. (2005) provides evidence for the existence of two broad political 'phenotypes' (observable characteristics of individuals that are determined by both genes and environmental influences): 'absolutist' or conservative, who exhibit a suspicion of out-groups and seek in-group unity; and 'contextualist' or progressive, who exhibit relatively tolerant attitudes towards out-groups. On this basis, it is reasonable to assume that conservatives would display greater political homophily than liberals in their linking behaviour in the blogosphere. However, as noted in Section 5.1, there are several other drivers of assortative mixing in social networks that may 'mask' the true level of homophily. In particular, there is the issue of group size, and this is where a difference between offline and online social network analysis becomes apparent. With online populations (such as the population of political bloggers in the US) it is very difficult to know the different population shares, and hence controlling for group size is a challenge or indeed impossible.

While the data collected by Adamic and Glance (2005) provide a useful insight into political homophily in the blogosphere at a point in time, it does not cast any light on the fragmentation hypothesis. Temporal analysis of political homophily in the blogosphere may provide some insights into the question of whether the web is leading us towards participatory democracy or cyberbalkanisation. Hargittai et al. (2008) look at the online behaviour of A-list political bloggers and test whether the amount of cross-ideological linking among blogs has declined over time by constructing a measure of assortative mixing similar to the homogeneity index presented in Section 5.1. They found no support for the fragmentation or cyberbalkanisation hypothesis, at least over a 10-month period; however, it would be worthwhile testing this over a longer period of time, using, for example, historical web data from the Internet Archive (Section 4.3.2).

7.4 AN INTRODUCTION TO POWER LAWS (ADVANCED)

While there are examples of power laws in the real world, many offline phenomena are distributed according to the normal or Gaussian distribution (Box 7.6). So what do power laws on the web look like?[5]

BOX 7.6   POWER LAWS IN THE REAL WORLD

The following draws from Clauset (2011).

In the real world, many empirical values cluster around a typical value. For example, the average height of adult US males is 1.763 m. This average is useful since not many US males deviate too far from this average (even the largest deviations are within 2 standard deviations from the mean, in either direction). The height of adult US males can be characterised using the normal distribution (or 'bell curve').

However, not all distributions fit this pattern. For example, if we look at the populations of the 600 largest US cities in 2000, the average population was 165,719, while New York City and Los Angeles are outliers with

---

[5]Also see Adamic (2002) for an introduction to the mathematics of power laws.

Assume we have data on 1,000 websites and we want to characterise the distribution of inbound hyperlinks (inlinks) to these sites (to understand, for example, how equal the websites are in terms of visibility or prominence). Assume we found that 1 website had a thousand inlinks, 9 had a hundred inlinks, 90 had ten inlinks and 900 had only one inlink. An *indegree–rank plot* is shown in Figure 7.2. On the horizontal axis, websites have been arranged according to rank order: the website with a thousand inlinks is on the far left, to its right are the 9 sites with a hundred inlinks, then the 90 sites with ten inlinks, followed by the 900 sites with only one inlink. It is clear from this

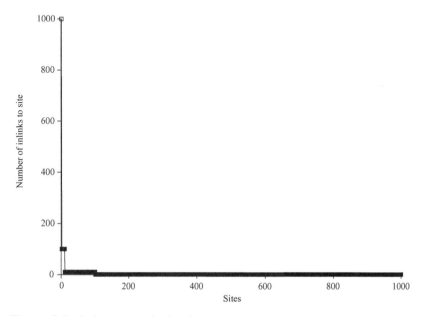

Figure 7.2 Indegree–rank plot demonstration data

figure that the distribution of hyperlinks is very unequal in that only a few sites receive a lot of hyperlinks and the vast majority receive only one hyperlink.

In this figure the Long Tail – the many sites that receive only one hyperlink – is clearly shown. In Table 7.1 the demonstration data are statistically summarised. The *cumulative distribution function*, CDF(x), gives the proportion of sites with indegree less than or equal to x: there are 900 sites with one inlink and so CDF(900) is 0.9. The CDF is plotted in Figure 7.3. The *complementary cumulative distribution function*, CCDF(x), is equal to $1 - \text{CDF}(x)$ and thus gives the proportion of sites with indegree *greater* than x, and this is plotted in Figure 7.4.

On the right-hand side of Table 7.1, logs (to base 10) of x and CCDF(x) are taken ($\varepsilon = 0.000\,000\,1$ has been added to the CCDF(x)

Table 7.1 Summary of power law example data

Number of inlinks (x)	Number of sites with indegree $\leq x$	CDF(x)	1–CDF(x)	$\log_{10}(x)$	$\log_{10}(1\text{–CDF}(x)+\varepsilon)$
1	900	0.900	0.100	0	−1.000
10	90	0.990	0.010	1	−1.996
100	9	0.999	0.001	2	−2.959
1000	1	1.000	0.000	3	−4.000
	1000				

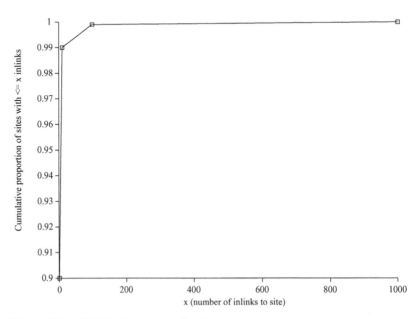

Figure 7.3 CDF for demonstration data

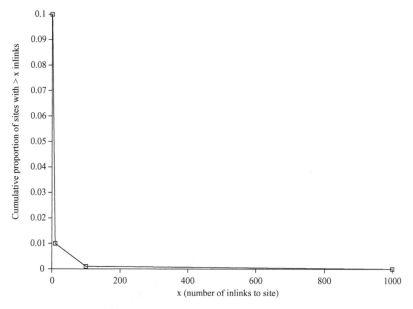

Figure 7.4　1 – CDF for demonstration data

because we cannot take the logarithm of zero). Finally, a log–log (or double-log) plot of CCDF(x) is shown in Figure 7.5; the straight line of this plot is the classic sign of a power law distribution (although,

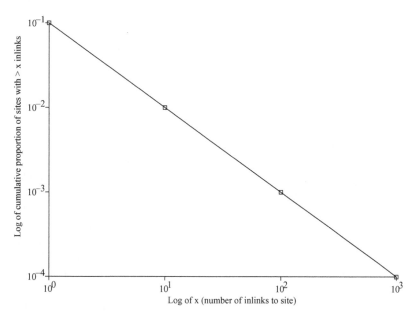

Figure 7.5　1 – CDF for demonstration data, log–log plot

typically, statistical methods are used to detect power laws, rather than just 'eyeballing' plots).

So, can we detect power laws in actual web data? Figure 7.6 shows the log plot of the CCDF for a hyperlink network of 646,720 Australian websites from a 2005 web crawl conducted by the Australian Commonwealth Scientific and Industrial Research Organisation (CSIRO). The straight line of this plot is indicative of a power law.

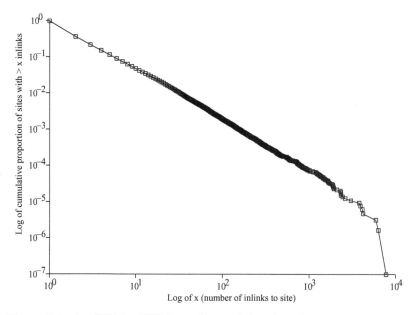

Figure 7.6　1 – CDF for 2005 Australian web, log–log plot

7.5 CONCLUSION

This chapter has looked at the evidence of the impact of the web on politics and society more generally. Regarding politics, we looked at the role of the web in promoting a level playing field. While the web is a low-cost medium for producing political messages, it is generally recognised that the 'build it and they will come' attitude does not work in an online environment that is increasingly competitive for the attention of web users. Research on political party use of the web has provided evidence that there has been a normalising of political power online, with major parties catching up on any ground that may have been lost to minor and fringe parties in the early days of the web. There is also research suggesting that the 'democratising' potential of the web with respect to the distribution of political information could be compromised by the existence of power laws in political hyperlink networks, which may promote a marked concentration of political attention.

The chapter also looked at how web use may be changing the political behaviour of individuals, and we assessed the evidence on the impact of web use on political engagement and also whether the narrowcasting nature of the web might be leading to political fragmentation or cyberbalkanisation. The effect of Internet use on social connectivity was also examined.

Further reading

See Hindman (2009) for more on the web and democracy, and Barabási (2002) for more on network science and power laws.

8

Government and Public Policy

This chapter is focused on government and public policy in the digital era. In line with the overall themes of the book, we are interested in how the web is affecting the operation of government and how web data are being used for public policy research.

The role of government in delivering information to the public is addressed in Section 8.1, while government authority is looked at in Section 8.2. Section 8.3 is focused on the use of virtual worlds for public policy modelling.

8.1 DELIVERY OF INFORMATION TO CITIZENS

e-Government refers to the use of new information and communication technologies (ICTs), and particularly the Internet, in order to achieve better government. This section is focused on how the web is affecting the delivery or provision of information to citizens.

Tools of government

The 'tools of government' analytical framework was first developed by Hood (1983) and was updated to the digital era by Hood and Margetts (2007). This framework contends that any public policy response will involve a mixture of four basic tools or resources: *nodality* refers to the property of being central in social and information networks, and thus having the capacity to deliver and collect information; *authority* refers to the legal power to forbid, demand, guarantee or adjudicate; *treasure* refers to control of taxation and the money supply; and *organisational capacity* refers to the government's stock of human and physical capital.

Margetts (2009) explains the concept of tools of government using the example of a policy response to smoking, which could involve public information and advertising about the health impacts (nodality), banning of smoking in public places (authority), free or subsidised nicotine alternatives (treasure) and the provision of counselling services to assist people to give up smoking (organisational capacity).

Government nodality is examined in the present section, while government authority is addressed in Section 8.2, and Box 8.1 is relevant to the treasure tool of government.[1]

BOX 8.1   PEER-PRODUCED DIGITAL CURRENCY – BITCOIN

The ability to control the money supply is one of the four 'tools of government' (Hood, 1983; Hood and Margetts, 2007). Bitcoin is a digital currency that can be traded for real-world goods and services. It is peer-produced (Section 9.1.1) in that the creation and issuing of Bitcoins, as well as the verification of Bitcoin transactions, are undertaken using a distributed peer-to-peer network of computers, with no involvement of government or a trusted third party.

There have been previous attempts to create digital currencies, but one of the innovations of Bitcoin, which was created by an anonymous cryptographer known by the pseudonym 'Satoshi Nakamoto', was a way of solving the so-called 'double-spending problem' – if a digital currency is just information (bits) and replicated costlessly, what is stopping someone from spending it as many times as they want? The conventional solution is to use a central clearinghouse to maintain a ledger of all transactions, thus preventing fraud, but the Bitcoin approach does away with the need for a trusted single third party by distributing the ledger across all computers participating in the peer-to-peer network.

This distributed nature of Bitcoin has led to it being known as a 'censorship-resistant currency', and the Wikileaks organisation turned to Bitcoin donations in 2011 when the US government put pressure on credit card companies to stop facilitating donations to the organisation. Bitcoins can be exchanged for real-world currencies and the exchange rate hit a high of nearly US$30 in June 2011 (but has since crashed to around $5). For more information on Bitcoin, see Barok (2011) and Wallace (2011).

8.1.1 Government hyperlink networks

In this section we look at two examples of research where hyperlink data have been used to shed light on the structure of government online and the provision of information to citizens.

---

[1]While organisational capacity is an equally important tool of government, and has been significantly influenced by the Internet (see Margetts, 2009), it is outside the scope of this book.

Hyperlinks and nodality

Government competes with other organisations (e.g. private or non-profit) in the delivery of services and information to citizens. Hyperlink data can be used to analyse nodality, or the relative position of government in social and information networks, which 'equips government with a strategic position from which to dispense information' (Hood, 1983, p.12). A government with strong nodality is better able to implement public policy without having to resort to the other tools of government (treasure, authority, organisation) which are typically more costly to wield, and strong nodality will help to maximise the effectiveness of the other tools (Escher et al., 2006, p. 4).

Our focus here is on delivery of information rather than delivery of service transactions (e.g. payments, registrations).[2] It is possible to collect anecdotal information on government nodality in the context of particular events; Escher et al. (2006) cite the example of government information websites performing poorly relative to those of NGOs such as the International Red Cross and private sector travel companies such as Lonely Planet in the aftermath of the 2004 tsunami in south-east Asia.

However, for the concept of nodality to provide useful and actionable insights to government, there needs to be a way of collecting data that can be used in comparisons across sectors of government and over time. As noted by Escher et al. (2006), before the advent of the web and the movement of government information services online, nodality was almost impossible to quantify, but the use of web crawlers to collect large-scale comparative datasets does provide a means of measuring it. But it needs to be recognised that ease of data collection is not enough: hyperlink data also need to have construct validity (Section 1.5) in the context of measuring government nodality.

Building on earlier work by Petricek et al. (2006), Escher et al. (2006) propose five measures of nodality that can be constructed using hyperlink data and website user experience data:

- *Visibility* – how likely is the website to be returned by search engines? Measured using counts of inlinks with the rationale that high counts of inlinks will raise the site's visibility in search engines (see also Section 7.1.1).
- *Accessibility* – once the website is found, how easy is it to find the required information? Measured using the percentage of inlinks that point to pages within the site (i.e. not the homepage), also referred to as 'deep links'. Deep links are more useful to users, since they take the user directly to the information, rather than having to navigate there from the homepage.
- *Navigation* – how easy is it to move around the site and find related information? Measured using: (1) the percentage of web pages that are in the

---

[2]Escher et al. (2006) note that information delivery involves just nodality, while delivery of service transactions is more about the other three tools of government – authority, treasure and organisation.

strongly connected component (Section 3.5), where a higher percentage is better, since all pages in the strongly connected component are reachable from each other; (2) the average shortest path length between web pages, with smaller values indicating that it is easier to follow internal links from one page to another; (3) user navigation experience (measured using laboratory experiments).

- *Extroversion* – to what extent are users directed to external providers of information? Measured using counts of outlinks – an extroverted site leads citizens to a wider range of information, hence increasing engagement.
- *Competitiveness* – how does the site compare with other similar sites? Measured by user experience (using laboratory experiments).

The authors used a purpose-built web crawler and user experiments to compute nodality scores for Australia, the UK and the US, focusing on a single government agency (the foreign office or department of foreign affairs). With regard to the user experiments, the authors compared the performance of the three foreign office websites by instructing users to find particular information under three 'treatment' conditions: (1) open access to the whole web; (2) access to the foreign office site only (with searching allowed); (3) access to the foreign office site only (no searching permitted).

The UK performed well on all but two of the nodality measures (accessibility and competitiveness), where it received a low score. In contrast, the US scored low on all measures except extroversion, for which it received a middle score. The Australian department of foreign affairs website was a 'mixed bag', scoring high on accessibility and competitiveness, medium on visibility and navigability, and low on extroversion.

Hyperlink structure of the .gov domain

Whalen (2011) analyses the hyperlink network of US federal government department websites with the aim of assessing the extent to which this structure mirrors the offline institutional structure of US government, and that of the overall web. The study is significant in its 'macro-level' approach (taking the entire structure of the entire .gov domain into account), which is a departure from earlier studies that focused on particular government websites (e.g. Petricek et al., 2006; Escher et al., 2006).

Whalen (2011) used LexiURL Searcher (Section 4.3.1) to construct a hyperlink network of 1,077 federal government websites. The websites were discovered using a snowballing technique, with the initial seeds being the websites of 85 US government institutions and agencies listed in the US Government Manual's organisational chart. In order to test whether the structure of the .gov domain reflects the offline hierarchical structure of the federal government, the author constructed the four graph-theoretic dimensions of hierarchy proposed by Krackhardt (1994). These four measures are used to quantify the extent to which a network is structured as an 'out-tree' (where all nodes are connected and all have indegree of one, except for the 'boss node' which is at the root of the tree):[3]

- *Connectedness* – this is the same as network inclusiveness (Section 3.5.2). If all nodes are in a single connected component, this indicates a 'unitary' structure, which is necessary for hierarchy.
- *Hierarchy* – the extent to which ties are reciprocated (reciprocal relations imply equal status, which is not the case in a hierarchy).
- *Efficiency* – the extent to which the network exhibits a tree-like structure, with each node having only one 'boss' or source of inlinks (this is termed efficiency because structures with multiple bosses involve redundancy with respect to the flow of information).
- *Least upper boundedness* – the extent to which each pair of nodes are downstream from a third common node (allowing a quantification of the unity of command in organisations).

Whalen (2011) computed the graph-theoretic dimensions both for the entire .gov domain and for subgraphs for 18 federal agencies, and concluded that, 'apart from moderate levels of reciprocation, the .gov domain is structured as a hierarchical tree with centralized sites surrounded by tiers of subordinate sites'.

The author also computed a 'silo' score for each agency subgraph, computed as the ratio of the total count of outbound links to the count of outbound links to other sites within the subgraph. Whalen (2011) proposes that this score can be used as a measure of the efficacy of provision of information to citizens (by forming lateral links across the greater .gov graph), and it ranged from 0.03 for Veterans' Affairs (high efficacy) to 4.6 for the Department of Defense (low efficacy).

8.2 GOVERNMENT AUTHORITY

The Internet has affected the ability of government to wield authority, by providing citizens with new means of challenging or circumventing authority, and government agencies with related technologies that can be used to monitor and respond (Margetts, 2009, p. 7). Section 8.2.1 looks at the role of the web in civil unrest, a topic that has much currency in the wake of the 2009 Iranian demonstrations, the Arab Spring and the Occupy Movement, all of which involved protesters making use of social media tools (in particular, Twitter). Section 8.2.2 looks at one aspect of the governance of the Internet itself where some governments' attempts to impose authority are being challenged: Internet censorship and content filtering.

---

[3]The following draws from Hanneman and Riddle (2005).

8.2.1 Civil unrest

As with many aspects of the social impact of the Internet, the role of social media in promoting social unrest is contested. This is best exemplified in the context of the role of social media in the Arab Spring. On the one hand, Gladwell (2011) has downplayed the role of social media in social unrest in the Arab Spring, stating: 'People protested and brought down governments before Facebook was invented. They did it before the Internet came along.' People on the other side of the debate see clear links between social media and civil unrest: 'Social media played a central role in shaping political debates in the Arab Spring. A spike in online revolutionary conversations often preceded major events on the ground. Social media helped spread democratic ideas across international borders' (Howard et al., 2011, p. 2).

There are two important questions here that social science can help answer. First, do social media change the probability of a given individual participating in social unrest? Answering this is difficult because establishing the counterfactual is probably impossible, but the question of whether the Internet is a shaping force or social tool is germane here (Section 1.6). Sociology does provide some insights into the question of whether social media use may influence participation in civil unrest. The issue here is whether social media can enable social ties that lead to the transmission of attitudes and behaviour (Section 5.2). As argued by Gladwell (2010), a critical point is whether the participation involves risk-taking. While the spread of information is aided by weak ties which potentially connect socially distant locations in the network (Box 3.3), participation in risky protest activity is an example of a 'complex contagion' which requires social affirmation from multiple sources (Centola and Macy, 2007).

If social media use leads to multiple sources of confirmation then this could change individual protest behaviour. However, Gladwell (2010) regards engagement in high-risk protest activity as requiring influence via strong ties, and he regards social media as mainly facilitating weak ties. Gladwell (2010) also comments on strengths and weaknesses of different modes of organisation (or governance mechanisms) (Section 9.1.1) in the context of social unrest, arguing that hierarchies are more likely to lead to the strong bonds and personal commitment that are required for successful (high-risk) protest activity than networked modes of organisation.

The second question is: given individuals have decided to participate in social unrest (e.g. public demonstrations or even rioting), how does the presence or absence of social media affect the dynamics of protest activities? This question is again very topical since the response of Egypt during the Arab Spring was to effectively 'turn off' the Internet in January 2011 in an attempt to remove an important medium of communication used by the protesters, and there were similarly calls to turn off the Blackberry network during the UK riots in 2011.

Again, it is very difficult to establish the impact of social media on the dynamics or operation of protest because of the difficulty of constructing the counterfactual (how would the UK riots and Arab Spring have turned out if there had been no access to social media?). One approach that may provide some insights is that of agent-based modelling, which is a computer simulation technique that can be used to model the emergence of complex phenomena arising from the actions and interaction of many agents (e.g. simulated individuals, organisations) who are programmed to display simple but socially realistic behaviour. Casilli and Tubaro (2011) use agent-based modelling to show that the decision to regulate or restrict social media in situations of civil unrest counterintuitively results in higher levels of violence. They find that a 'complete absence of censorship ... not only corresponds to lower levels of violence over time, but allows for significant periods of social peace after each outburst' (p. 2).

8.2.2 Internet censorship

Section 1.3.1 looked at the cyberspace ethos that developed during the pioneering years of the Internet. One important aspect of this ethos relates to freedoms. There is an expectation that there should be greater freedom in cyberspace than in the real world, and consequently there is resentment of any activity by governments and corporations aimed at constraining those freedoms. This expectation is famously captured in John Perry Barlow's 1996 Declaration of the Independence of Cyberspace (Section 1.3.1). Internet censorship is one of the main areas where these assumed entitlements to freedom are challenged (the other is in the area of copyright).

While it is well known that China implements strict Internet filtering via the 'Great Firewall of China', plans of the Australian and UK governments in this area have raised concerns amongst academics and activists. The position of the Australian government on Internet filtering is contained in the Australian Communications and Media Authority (ACMA) February 2008 report *Developments in Internet Filtering Technologies and Other Measures for Promoting Online Safety*. The UK's House of Commons Culture, Media and Sport Committee released in July 2008 a report titled *Harmful Content on the Internet and in Video Games*.

There is surprisingly little social science academic research on the topic of Internet censorship.[4] Further, a lot of the research that is focused on Internet censorship takes a normative stance (in line with the cyberspace ethos of freedom) that any Internet censorship is bad, with research focusing on identifying the countries most active in censorship. The Open Net Initiative (ONI)[5] is described in the following way on its homepage: 'Internet censorship and surveillance are growing global phenomena. ONI's mission is to identify and document Internet filtering and surveillance, and

----

[4]Depken (2005) reviews social science research on censorship more generally; see also Lambe (2004) for research into attitudes towards censorship.

[5]http://opennet.net

to promote and inform wider public dialogs about such practices.' ONI undertakes several activities, including: developing software tools for enumerating the extent of Internet filtering and surveillance, and for circumventing Internet censorship by allowing users to access blocked web pages; capacity building among activists and researchers in countries where Internet censorship is prevalent; and academic research into the extent and trends of Internet filtering and surveillance.

So what do the public think about Internet censorship? In 1999 the Australian Broadcasting Authority (the predecessor of ACMA) released details of an international comparison (Australia, Germany and the US) of attitudes to Internet censorship (Kocher, 2000). Of the 1,000 Australians in the sample, 60% said they favoured the blocking of objectionable content on the Internet, although 77% said that content-policing measures were likely to be ineffective. At the time of the introduction of the Australian Broadcasting Services Amendment (Online Services) Act, which brought online content into line with other Australian mass media with regard to censorship, there was debate about the level of public support for Internet censorship. Online survey data 'showed limited support for government censorship among Internet users – a measure of concern, rather than the perception of a social problem' (Chen, 2000).

The demand for Internet censorship

Depken (2005) uses an economic model to provide a rationale for why an individual might support Internet censorship, even if censorship is expected to reduce the amount of 'acceptable' information that is available (this is what ONI refer to as 'overblocking'[6]). The author then conducted an empirical analysis of the demand for censorship using an online survey of 5,022 people from 1998, which included information on personal characteristics, experience and ease with using the Internet, and attitudes towards censorship. Relative attitude towards censorship was measured using the following question: 'Please indicate your agreement/disagreement with the following statement. I believe that certain information should not be published on the Internet', with answers being reported on a scale of 1 to 5 (1 being 'agree strongly' and 5 being 'disagree strongly').

Depken (2005) acknowledges the potential problem of sampling bias – because the survey was administered online, only those with Internet access were able to participate. One might expect that those with access to the Internet are less likely to favour Internet censorship (although there is no theoretical reason for this to go either way).

It was found that the following characteristics were associated with a greater demand for censorship: having children, being married, being older, using the Internet for religious content, and working in the public sector.

---

[6]http://opennet.net/about-filtering

Those who were male, living in urban areas, used the Internet for political reasons, had more Internet experience and were more comfortable on the Internet tended to be against Internet censorship.

8.3 PUBLIC POLICY MODELLING

In this section we first look at when virtual worlds can be used for modelling public policy, and then provide an example of online natural and laboratory experiments in the context of public policy design.

8.3.1 The mapping principle

Williams (2010) contends that there are two types of research into virtual worlds. The first focuses on understanding who is in these spaces and why, and characterising the types of computer-mediated communication that are taking place. The second type of research into virtual worlds is based on the premise that behaviour in virtual worlds parallels that in the offline world (i.e. there is a 'mapping' between the two worlds), and hence an understanding of human behaviour in virtual worlds can provide insights into human behaviour in real life.

Williams regards this second type of scholarship into virtual worlds as 'more radical', and argues that if there is a mapping, virtual worlds can be used as a virtual 'Petri dish' for modelling human behaviour, and have immense potential for social science research. However, Williams argues that this mapping cannot be assumed: we need to know when and where we can leverage virtual worlds to learn about human behaviour.

He illustrates the problems of assuming there is a mapping between virtual worlds and the real world by reference to the 'World of Warcraft Plague'. A programming error meant an infectious deadly 'disease' designed to be active only in a small geographic area (accessible by advanced characters who could withstand the 'health' damage) spread to other parts of the world, killing many characters. This appeared to provide a research opportunity; epidemiologists published on it in leading journals (Balicer, 2007; Lofgren and Fefferman, 2007) and terrorism researchers saw an opportunity to understand what people do when infected, and how the flow of information was affected by or contributed to the outbreak.

But Williams is sceptical of the use of virtual worlds for epidemiological research. While there might have been some emotional pain associated with having your character killed off there is no physical effect on the player, and this led to behaviour that was not realistic (characters were seen chasing friends trying to infect them, for example). This led Williams to conclude that WoW does not provide risks and rewards that allow a valid modelling of the impact of disease outbreak – that is, there was not a mapping in this case.

The testing of the mapping between the online and offline world is one of the major themes in this book. Section 9.3.2 looks at the work of Burt

(2011) on whether virtual world data have construct validity (Section 1.5) in the context of research into structural holes and performance. In Section 5.1 we looked at whether one of the major social forces that exists in real-world friendships, homophily, also exists in online friendships. Section 9.2.1 looks at the construct validity of hyperlinks as measures of scientific output and impact.[7]

8.3.2 The macroeconomics of a virtual world

Castronova et al. (2009) use data from the virtual world EverQuest 2 (300 million individual transactions on thousands of items sold, with purchase amounts and prices) to construct three macroeconomic indicators (GDP, price level, money supply) in order to test whether there is a 'mapping' on the basis of macroeconomic behaviour from virtual worlds to the real world. The authors argue that if these macroeconomic aggregates move in similar ways to their offline counterparts, then this greatly enhances the viability of virtual worlds for economic policy-related research. As they put it: 'If virtual worlds are at all similar to real ones, then the core elements of their economies should look and work like real ones' (p. 691).

In order to construct macroeconomic measures that resemble the offline counterparts, the authors focused on goods sold by one player to another via the EQ2 consignment system, which is similar to broker trading in a real economy, and constituted around 7% of the total transactions in EQ2 over the period studied. They found evidence that the quantity theory of money (which predicts that there is a direct and proportional relationship between the price level and money supply) holds in EQ2. However, they also found that the macroeconomic aggregates fluctuated much more markedly than they do in the real world, and conclude: 'Taken altogether, this is evidence that virtual economies are not perfect analogs of real economies at the aggregate level. They seem to be less stable. ... Perhaps virtual economies are a very precise analog for other kinds of real-world economies, such as frontier, developing or black market economies' (p. 71).

8.4 CONCLUSION

This chapter has considered how the web has influenced government and public policy, and how web data can be used for research in this area. First, we looked at two examples of research where government hyperlink data are used for quantitative analysis of government provision of information to

---

[7]See also Yee et al. (2007), who show that the social norms of gender, interpersonal distance and eye gaze transfer into virtual worlds, with the implication that 'it is possible to study social interaction in virtual environments and generalize them to social interaction in the real world' (p. 119).

citizens. Next, we examined the impact of the web on government authority from the perspective of the role of social media in civil unrest and
government-enacted Internet filtering. Finally, we considered the use of
virtual worlds for public-policy-related research.

Further reading

See Hood and Margetts (2007) for more on the tools of government in the
digital age. Deibert et al. (2008) document the extensive research by the
OpenNet Initiative on the trends and patterns of Internet filtering. For
more information on the use of virtual worlds for social science research,
see Castronova (2007, 2009).

9

Production and Collaboration

This chapter looks at how the Internet has changed the production of information-based goods. It also considers how web data can be used for measuring production and impact in the academic sector, and whether the web may have led to a reconfiguring of access to scholarly information. Finally, the chapter shows how data from virtual worlds can be used to provide new insights into the contribution of social networks to individual achievement.

Section 9.1 introduces peer production, which is a special case of information good production, and looks at the motives for contributing to peer production. Section 9.2 looks at how the web has provided new opportunities for quantitative measures of academic output and also discusses the potential effect of the web on the distribution of academic visibility and authority. Section 9.3 discusses the difficulties of empirically identifying the influence of social network structure on individual achievement, and shows how data from virtual worlds may shed new insights into structural holes and achievement.

9.1 PEER PRODUCTION AND INFORMATION PUBLIC GOODS

Section 6.2.1 introduced the concept of an information public good, which is an information-based good (e.g. software, databases, encyclopedias and repositories) that exhibits the characteristics of non-rivalry and non-excludability. The example of an information good in Section 6.2.1 was a hyperlink network formed by NGOs; while the NGOs were not hyper-linking with the conscious objective of creating an information public good, this was the result of their distributed activities.

In contrast, in this section we introduce a class of information public goods which are created by the conscious or deliberate actions of people. This collaboration, which is typically geographically distributed and Internet-enabled, is known as *commons-based peer production* (see Benkler,

2006; Demil and Lecoq, 2006); however, we use the abbreviated term 'peer production' here. The section also looks at what motivates individuals to contribute to activities that result in the creation of information public goods, and presents an example of the production of information public goods.

9.1.1 Peer production

The best way of introducing peer production is to look at some well-known examples:

- Open source software is software where the users of the software have the right to modify and distribute the source code (Box 9.1). Open source projects are often contributed to by hundreds of programmers from around the world. Some of the best-known examples of open source software are the Linux operating system, the Apache web server, the MySQL relational database, and the Perl and PHP programming languages.
- Wikipedia (Section 1.2).

BOX 9.1   WHAT IS OPEN SOURCE SOFTWARE?

Software generally falls into two categories: *compiled* or *interpreted*. With compiled software, the source code (this is the 'source' in 'open source'), which is human-readable (this is what the programmers write), is converted by a piece of software called a compiler into a set of instructions that can be understood by the computer (the compiled code is called the 'binary' or 'executable'). In contrast, software written in an interpreted language remains in human-readable form (it gets converted into machine-readable form when it is run, i.e. at 'run time'). Examples of compiled software languages are C and C++, while Perl and PHP are examples of interpreted software languages.

The traditional model for software development (i.e. the model followed by companies such as Microsoft) has been to only distribute the executable files. If you want to use Excel or Word, you do not need the source code – all you need is the executable or binary. However, proponents of open source argue that users of software should be given the rights (via the software licence) to make use of the source code as well as the executable files. The various open source licences of the Free Software Foundation (such as GNU GPL or GNU LGPL) give people the right to copy, modify and distribute freely the source code, with the proviso that any 'derived work' (new software created via the modification of existing source code) must be covered by the same open source licence. That is, it is not possible under an open source licence to take open source software, modify it, and then sell it as closed source software (i.e. without the source code, or with more restrictive licences on the source code).

- NASA Clickworkers[1] was a site that allowed ordinary citizens to contribute to a US National Aeronautics and Space Administration project classifying craters on Mars according to size and age.
- SETI@home[2] is a scientific experiment where ordinary citizens can participate in the Search for Extraterrestrial Intelligence (SETI) project by running a free screensaver program that downloads and analyses radio telescope data.

Hierarchies, markets and the bazaar

In order to understand what peer production is, it is helpful to learn about the organisational mechanism or governance structure that underpins it.

In economics, a *transaction cost* is the cost of making an economic exchange. The transaction cost is any cost over and above the price of the item being purchased. When you purchase shares the transaction cost is the brokerage cost, plus any other costs you may have incurred (e.g. you may have paid for market advice). Formally, transaction costs include (Dahlman, 1979): search and information costs (the cost of finding the good that you need, at the lowest price); bargaining and decision costs (the cost of agreeing on the terms of the transaction, e.g. the price); and policing and enforcement costs (the cost of ensuring that all parties to the transaction do what they said they will, and taking appropriate action if not).

A *governance structure* is an organising mechanism that is designed to reduce transaction costs, and involves a *contractual framework* and *control and incentive mechanisms*. The governance structure that underpins peer production is known as 'bazaar governance', and Demil and Lecoq (2006) argue that the uniqueness of bazaar governance is best shown by comparing it with other established governance mechanisms (Table 9.1).[3]

A good example of a *hierarchy* is a firm. In a firm, the exchange of labour is controlled by employment contracts (this is the contractual framework).

Table 9.1 Comparing governance structures: market, hierarchy and bazaar (Demil and Lecoq, 2006)

	Market	Hierarchy	Bazaar
Contractual framework	Classical contract	Employment contract	Open licence contract
Incentives intensity	High	Low	Low
Control intensity	Low	High	Low

[1]http://clickworkers.arc.nasa.gov

[2]http://setiathome.berkeley.edu

[3]Demil and Lecoq (2006) included a fourth governance mechanism in their analysis (network), but we do not cover that here.

Control mechanisms are typically strong (you have to do what your boss says) and incentive mechanisms are generally quite weak (only particular industries, e.g. finance, have strong performance-related incentives in the form of bonuses). *Markets* are supported by classical contract law – there is strict adherence to the contract terms (e.g. relating to the description of a good or service), and disputes are resolved by the courts. Markets are characterised by strong incentive mechanisms and relatively weak control mechanisms (the definition of a properly functioning market is one where no party can exert undue control).

Bazaar governance is supported by the 'open licence contract' (Box 9.1). With bazaar governance the incentive mechanisms are weak (the people who contributed to the NASA Clickworkers project, for example, received no pay) and the control mechanisms are also weak (in a typical open source software project no one is obliged to perform any particular task, and work is not imposed or mandated by a leader).

9.1.2 Information public goods

In this section we look at what motivates people to contribute to the creation of information public goods. We also look at empirical techniques for characterising people who are 'good contributors' to the creation of information public goods that arise from threaded conversations in online forums and newsgroups.

Motivations for contributing to creation of information public goods

We have seen that one of the hallmarks of peer production is that both the incentive and control intensities tend to be low. Why, then, do people contribute to peer production? With regard to information public-goods creation more generally, the incentives for contributing to the creation of some information public goods (e.g. being the 'answer person' in a Usenet group – see below) are possibly even lower than with peer production. What motivates people to contribute to the creation of information public goods?

The following discussion focuses on the case of open source software, but many of the ideas are applicable to other examples of peer production, and information public goods more generally. The motivations for participation in an open source software project can be classified into two broad types: extrinsic and intrinsic.

Extrinsic motivation comes from factors outside an individual – this is where the person doing the work is motivated by some kind of reward (e.g. money) and not purely by an enjoyment of doing the task or some other reason. Not surprisingly, it has been economists who have promoted the view that contributors to open source projects are motivated by extrinsic rewards. Specifically, it has been argued that participating in an open source project is an example of labour market *signalling* in that the programmer can develop a reputation as a good coder that will lead to job opportunities or higher wages (Lerner and Tirole, 2002).

Intrinsic motivation is used to explain why a person engages in an activity when there are no obvious external incentives. With open source software development, the following intrinsic motivating factors have been suggested: the developers contribute to software that they want to use themselves (this is the user-programmer phenomenon); developers get enjoyment out of writing code ('coding is fun'); and 'gift culture'. The third intrinsic factor (gift culture) is probably the most important and involves several potential motivations: giving code to the community might help to establish the programmers' 'social status'; 'community identification' (programmers identify with the open source movement or community, which has shared goals, e.g. opposition to proprietary software such as 'Micro$oft'); and altruism (a motivation to do something for others with no expectation of reward).

Identifying the 'answer person' in Usenet groups

Welser et al. (2007) use data from the Microsoft Netscan Project (Section 3.3.2) to develop a statistical approach for identifying 'answer people', individuals who mainly respond to questions posed by other users instead of posing their own questions, or getting involved in unproductive 'flame wars'.[4]

Why is it important to identify answer people and encourage their behaviour within newsgroups and forums? Welser et al. (2007) argue that Usenet, as a searchable repository of answers to questions on just about any topic, is an example of an information public good. Usenet is non-rival since my action of using Google Groups to find an answer to what to feed my cat does not affect anyone else's ability to search Usenet for information. Usenet is non-excludable since anyone with an Internet connection can 'consume' information contained within it. To the extent that Usenet can be viewed as an information public good, answer people are important since they are the major contributors to this public good. Being able to identify answer people is important since it will allow the development of better approaches for cultivating communities of practice where such sharing is prevalent.

Welser et al. (2007) focus on three Usenet newsgroups: comp.soft-sys. matlab (statistical/mathematical software), microsoft.public.windows.server. general (Microsoft server) and rec.kites. They propose two structural signatures associated with answer people that can be derived from the threaded conversation network data.

First, answer people tend to participate in threads initiated by others and typically only contribute one or two messages per thread. The authors show

---

[4]See also Turner et al. (2005) and Smith et al. (2007).

this visually via the use of 'authorlines' (Figures 9.1 and 9.2) which are a visual representation of a person's contribution to a newsgroup. In an authorline, each time point represents one week; the circles above the horizontal axis reflect a person's contributions to threads they initiated, while responses to a thread started by someone else are shown by the circles below the horizontal axis. The size of the circles is proportional to the number of messages the author contributed to each thread. An exemplary authorline for an answer person is displayed in Figure 9.1, while that of a discussion person is in Figure 9.2.

Second, the ego networks of answer people tend to contain alters who themselves answer few, if any, questions posed by others (Figure 9.3). Further, the 1.5-degree ego network of answer people tends to have small proportions of triads (i.e. their neighbours are not neighbours of each other) and they have few intense ties (i.e. they seldom send multiple messages to the same recipient). In contrast, the ego network of a discussion person contains highly connected alters, and there are many intense ties (Figure 9.4).

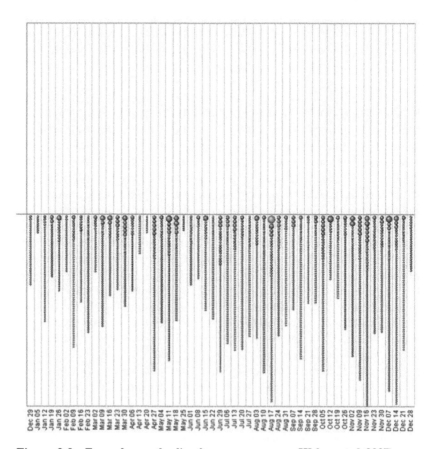

Figure 9.1 Exemplary authorline for answer person, (Welser et al. 2007)

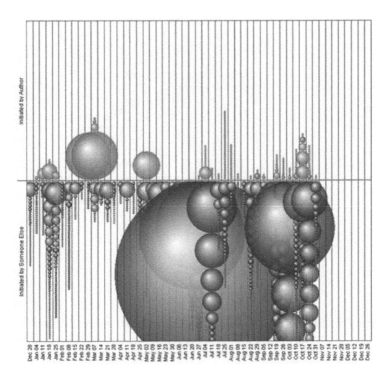

Figure 9.2 Exemplary authorline for discussion person, (Welser et al., 2007)

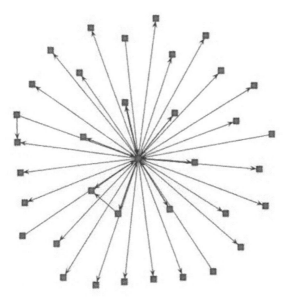

Figure 9.3 Exemplary ego network for answer person, (Welser et al., 2007)

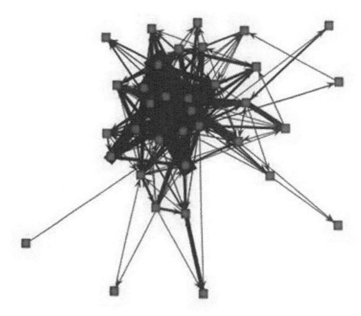

Figure 9.4 Exemplary ego network for discussion person, (Welser et al., 2007)

9.2 SCHOLARLY ACTIVITY AND COMMUNICATION

This section looks at how web data can be used for constructing metrics of scientific production and whether the web may be leading to a reconfiguration of access to scholarly authority and expertise.

9.2.1 Webometric measures of scholarly output and impact

While the term 'webometrics' (see Section 4.2.1) is sometimes used to refer to any quantitative analysis of web data, many applications of webometrics have been in the context of scholarly output and activity, and the field has its disciplinary origins is informetrics (citation analysis is an example of informetrics).

An example of webometric research is that of Barjak and Thelwall (2008), who set out to identify the factors associated with the prominence or visibility of websites belonging to research teams in the life sciences, in the context of assessing the viability of hyperlinks as scientometric performance indicators. This process of assessing the viability of hyperlinks as science and technology performance indicators can be seen as one of construct validation of hyperlink data in this context (Section 1.5).[5] Academic research performance is a real-world 'construct' which is measured using various qualitative and quantitative data, for example peer review and bibliometric measures such as number of publications and citation counts. Evaluation of the construct validity of hyperlink data involves being able to show that hyperlink counts are correlated with other observable variables that are known to correlate with academic performance. If hyperlinks can be validated as measures of scientific performance then the 'World Wide Web [can] be added as an additional database for the description and analysis of scientific processes and structures' (Barjak and Thelwall, 2008, p. 628).

Barjak and Thelwall (2008) collected data on over 400 research teams in the life sciences from over 10 European countries, and employed hyperlink-counts regression using the count of inlinks to the websites of the research teams (found via search engines) and the characteristics of the website and the research teams. The authors found that only the size of the website and the size of the team were significantly correlated with inlink count (country of origin was also significant, but appeared to be related to search engine coverage). Other factors that have been shown to correlate with academic performance (publication output of the team, publication quality, extent of research collaboration with other teams, connections with industry, average team age, team leader sex and professional recognition) were not correlated with hyperlink inlink counts. The results also suggested that male team leaders and team leaders with professional recognition do not receive more inlinks as a result of these characteristics, but have larger teams and websites.

On the face of it, these findings are not great support for the construct validity of hyperlinks for scientometric analysis. However, some of these results were not consistent with the findings of other webometric research in this area, and the authors concluded that more research is needed. But they also noted that there are practical constraints on the use of hyperlink data for analysis of academic performance. First, the count data varied markedly across the three search engines they used (Google, Yahoo and MSN), and they recommended that multiple search engines be used in link analysis since the peculiarities of any given search engine might bias the results. Further, the analysis was complicated by the fact that hyperlink inlink counts are not normally distributed (as discussed in Section 7.1.1, there is evidence that the distribution of hyperlink counts follows a power law), and so standard regression techniques are not appropriate. Barjak and Thelwall (2008) concluded that hyperlinks 'are probably still impractical to use for routine Science and Technology indicators at the research group level' (p. 641).

---

[5]See also McNutt (2006), who argues that hyperlink data can proxy real-world structure in the context of policy networks, and Rogers (2010a), who proposes digital methods as a way of learning about the offline world using data generated online.

9.2.2 Reconfiguring access to scholarly information and expertise

A body of research has considered whether the web (or ICTs more generally) may be 'reconfiguring access' (Dutton, 2005) to scholarly information and expertise, hence altering the distribution of academic authority or prominence at the level of individual researchers, teams, universities and countries. This work thus mirrors political science research into the effect of the web on the distribution of political power (Section 7.1).

At the heart of this research is the question of whether the web can play a role in 'democratising' access to scientific expertise (Caldas et al., 2008). Drawing on the 'retrievability versus visibility' argument presented in Section 7.1.1 in the context of the distribution of political information, the advent of the web has led to individual researchers and research centres around the world being potentially equally *accessible*, but are they necessarily equally *visible*? That is, the project website of a research team in Mumbai is just as retrievable as that of their counterparts in Oxford, but will the work of the Mumbai researchers show up in search engines or be found via casual web surfing, and hence be as likely to be cited as that of the Oxford researchers?

We saw in Section 7.1.1 that web visibility is largely related to hyperlink structures, and so the question of whether the web can democratise access to scientific expertise involves asking two separate questions. First, will offline hierarchies in scientific expertise simply be reflected in hyperlink structures that benefit better-resourced research teams from wealthy countries, or does the web provide an opportunity for smaller and less well-funded research teams to 'leapfrog the competition'? This question is analogous to that asked in Section 7.1 in the context of politics and the web: does the web allow minor political parties to compete on equal footing with the better-resourced major parties (Box 7.1)?

To the extent that hyperlinks have 'construct validity' as measures of scientific output and authority (Section 9.2.1), hyperlink structures (and hence web visibility) should reflect offline hierarchies. While not focusing on hyperlinks specifically, Matzat (2004a) found that participation in academic Internet discussion groups does not lead to an equalisation of the relations between academics on the periphery of the field and those who are well established or central.

The second, related, question is: is there anything inherent in the web that could be contributing to academic expertise and authority becoming more or less concentrated? There is strong evidence for the existence of power laws on the web (Section 7.1.1). The process of preferential attachment (Barabási and Albert, 1999) that has been posited as contributing to the emergence of power laws is a process of cumulative advantage (also known as 'rich-get-richer', 'winner-takes-all' or the 'Matthew effect') which has also been observed in the context of the distribution of academic authority or prominence in science (Merton, 1968).

So the question is whether the web is likely to increase the Matthew effect, thus concentrating scientific expertise in the hands of a smaller number of 'superstar' scientists (Ackland, 2010a), or whether it will contribute to a reconfiguration that will lead to a more equal distribution of scientific expertise. There is not strong empirical evidence to evaluate this question, and this is largely because of the difficulty in constructing the counterfactual (what would have happened to the distribution of academic expertise in the absence of the web?). Caldas et al. (2008) present some preliminary findings which indicate that the web may have a democratising effect, but this is based on indirect evidence that the winner-takes-all phenomenon (i.e. power laws) may not be evident in the academic web. But the authors surmise that this might be because the web has contributed to a fragmentation of scientific fields, thus allowing the emergence of many winners (big fish in many small scientific ponds).

The research into how the web may be contributing to a reconfiguring of access to scholarly information and expertise also has clear parallels with research into recommender systems and sales diversity (Section 10.2.3). The above summary has focused on the *positive dimension* of the research, that is, how the web is potentially changing access to scholarly information and expertise. But there is also a *normative dimension*: to the extent that the web is contributing to greater (or lesser) concentration of scientific expertise, is this a good thing? Addressing the normative considerations of this debate is beyond the scope of the present text, but we can note that the use of the word 'democratizing' in the subtitle of the work by Caldas et al. (2008) suggests that the authors believe there should be a reduction in the concentration of scholarly expertise or authority. We are not aware of a formal normative treatment of the 'web and scholarly expertise' debate, but the economic welfare analysis of recommender systems in Fleder and Hosanagar (2007) might provide some insights (Section 10.2.3).

9.3 NETWORK STRUCTURE AND ACHIEVEMENT

While Section 5.2 looked at how socially connected individuals exert influence on each other (social influence), this section is about how the overall network position of an individual might influence outcomes and performance.

9.3.1 Identifying a 'network effect' in outcomes

There can be spurious correlation between networks and behaviour or outcomes when network structure has a predictable impact on a successful outcome and hence the agent has a strategic advantage in acquiring particular types of network ties. The following are three examples where this might happen:

- Innovation in firms – certain network structures are likely to influence the ability of the firm to acquire knowledge leading to innovation and performance improvement (e.g. Uzzi et al., 2007).
- Labour markets – Granovetter (1973) shows that weak ties are important in labour market outcomes.
- Occupational success – Burt (e.g. Burt, 1992) shows that the ability to fill structural holes in social networks positively affects occupational progress.

Given that in the three examples above there is a clear connection between network structure and outcomes, it is in the interest of the agents to strategically acquire ties. Some will succeed and some will fail, but those who succeed are likely to have improved performance. But in this situation, is it the network that is causing the outcome or is it an unobserved attribute of the agent that allowed it to successfully acquire the best network structure and to perform better than its peers? That is, is it really the network that is causing the outcome or is it an individual-level (perhaps unobservable) attribute such as talent or drive that is important?

The argument that networks cause outcomes can thus be undermined in such situations because of the possibility of *unobserved heterogeneity* (variation in network position across agents reflects intrinsic ability of the agents, which will also affect performance) and *reverse causality* (prior performance differences are responsible for current network position).

9.3.2 Structural holes in Second Life

In Section 8.3 we saw that when there is a 'mapping' (Williams, 2010) between online and offline behaviour, data from online environments can be used to provide new insights into human behaviour, for example in the context of public policy design. The sociologist Ronald Burt has conducted pioneering research into the social structure of competitive advantage, and has recently studied how network position affects performance in Second Life (Burt, 2011). This work is aligned with Section 8.3 in that Burt asks whether network processes in virtual worlds are the same as those in the real world, that is, whether virtual world networks have construct validity (Section 1.5):

Do social networks in virtual worlds have the same effects observed in the real world? The advantages of network data in virtual worlds are worthless without calibrating the analogy between real and inworld. If social networks in virtual worlds operate by unique processes unrelated to networks in the real world, then the scale and precision of data available on social networks in virtual worlds has no value for understanding relations in the real world. On the other hand, if social networks in virtual worlds operate just like networks in the real world, then we can use the richer data on virtual worlds to better understand ... network processes in the real world. (Burt, 2011)

The particular effect that Burt is focused on relates to his research into 'structural holes' (Box 3.4). Burt (2011) studies both brokerage and closure in Second Life, making use of three types of relational data: one-to-one friendships (which operate in a manner similar to Facebook), group membership (again, similar to that in Facebook) and rights granted to friends. With regard to the latter, Second Life allows a user to grant another user three levels of access (which imply increasing levels of trust): the right to know when you are online, the right to know your location in Second Life, and the right to directly modify your online inventory (thus influencing the appearance and behaviour of the avatar).

Burt constructed ego networks by defining and quantifying four types of directed tie: 'no tie' (users i and j are not friends); 'weak tie' (users i and j are friends, but i granted j either no rights or only the right to know when i is online); 'average tie' (i and j are friends, and i granted j the right to know i's location inworld); and 'strong tie' (i and j are friends, and i granted j the right to modify i's online inventory).

With regard to the closure hypothesis, there is no obvious Second Life analogue to team efficiency measures and so Burt focuses on the intermediate variable of trust and hypothesises that closure is associated with higher levels of trust in relationships. Trust was measured by the three levels of rights that friends can grant each other, and Burt found support for the closure hypothesis, showing that the level of trust between two users was positively related to the level of network closure around the friendship, measured by the number of indirect connections between ego and the friend.

With regard to brokerage, the hypothesis is that occupying structural holes in Second Life is correlated with achievement, which Burt measured as the contribution of ego to the formation and maintenance of groups (one of the main types of in-world infrastructure that attracts people to Second Life). It was found that people who have greater access to structural holes (measured by the number of non-redundant friends, i.e. friends who are not connected to any other of ego's friends) founded more groups, and their groups were more likely to be successful in terms of membership and activity.

It should be noted that Burt's results are based on a snapshot of activity within Second Life: no use is made of the rich timestamp data that allow one to know the exact sequence of friendship and group formation, and (as Burt notes) his analysis does not allow conclusions about causation. In particular, it may be that two people with a high level of trust in their relationship subsequently acquire many mutual friends (hence building network closure). Similarly, a person who has established successful groups in Second Life may acquire friendships from diverse parts of the virtual world, with access to structural holes thus resulting from achievement, rather than the other way around.

This problem of causation versus correlation in network analysis is typical of offline studies into structural holes, and making use of the dynamic network data (one of the strengths of data collected from online environments such as Second Life) will allow researchers to better understand the exact causal relationship between network position and outcomes.

9.4 CONCLUSION

This chapter has looked at production and collaboration in three different contexts. First, we revisited the topic of information public goods and also looked at peer production, which describes the creation of a particular class of information public good, where distributed teams of people collaborate via the Internet. We then turned to production in the academic sector, and looked at how hyperlink data can provide new insights into scientific productivity, and whether the web has brought about a reconfiguration of access to scholarly output and authority. Finally, we focused on the influence of networks on individual performance and provided an example where data from Second Life are used for conducting research into structural holes.

Further reading

For more on peer production, see Benkler (2006). The use of web data for measuring academic output and impact is covered in Thelwall (2004, 2009b), and contributors to Dutton and Jeffreys (2010) provide a thorough investigation into how the web is transforming academic processes in the sciences and humanities. Burt (1992) is the authoritative introduction to research into structural holes.

10

Commerce and Marketing

This chapter is focused on Internet commerce and marketing, looking at the extent to which the Internet has changed the way goods and services are marketed, and how web data are providing new insights into the role of social factors in people's decisions to buy or adopt products.

In Section 10.1 we introduce the concept of the Long Tail, which is the idea that the Internet has made it profitable for businesses to focus on selling niche products rather than relying on just the blockbusters. We review the empirical evidence as to whether businesses really are focusing their business strategy on niche products in the Long Tail. A central aspect of the Long Tail argument is that sales are becoming more diverse (i.e. consumers are spending more of their money on niche products compared to superstar products), and that this change in sales pattern has been brought about by the Internet. In Section 10.2 we look at research on how Internet ratings and recommender systems impact on consumer behaviour, and we also revisit the topic of social influence (Section 5.2), but this time in the context of word-of-mouth advertising.

10.1 DISTRIBUTION OF PRODUCT SALES

The *Long Tail* is the term used by Anderson (2004, 2006) to refer to an Internet marketing strategy that involves selling large numbers of unique or 'niche' items, with relatively small sales volumes for each item.[1] In this section we define a Long Tail (which relates to the concept of power laws, introduced in Section 7.1.1), use the 'economics of superstars' to see how a Long Tail might develop, and evaluate the relevant empirical evidence. While there is still academic debate about the existence of the Long Tail and its relevance for business, there is no doubt that this has been an influential business concept.

---

[1]Anderson was inspired by a web post by Clay Shirky titled 'Power Laws, Weblogs, and Inequality' (http://www.shirky.com/writings/powerlaw_weblog.html).

10.1.1 Power laws and superstars

A market (e.g. for music CDs, film DVDs, books, downloaded songs) can be described as having a Long Tail if a very small number of 'superstar' or 'blockbuster' products account for the lion's share of sales, while the vast majority of items are sold in only very small numbers.[2] We are therefore talking about a market where there exists a highly unequal distribution of sales, that is, where sales are distributed according to a power law (Section 7.1.1), or some other very unequal distribution.

To see how power laws and Long Tails are related, it is useful to revisit the demonstration data from Section 7.1.1, which consisted of 1,000 websites (these data do not have anything to do with Internet commerce, but will help us to understand Long Tails). In Figure 7.2 we presented the indegree–rank plot: websites were ordered along the horizontal axis according to indegree (number of inbound hyperlinks pointing to site), with indegree being measured on the vertical axis. Figure 7.2 is replicated in Figure 10.1, with the Long Tail shown (the vast majority of websites having very few inlinks).

In Section 7.1.1 we also showed that a power law distribution can be shown as a straight line in a double-log plot of the CCDF (Figure 7.5); this is the visual 'telltale sign' of a power law distribution. Another way to plot a distribution is to use the *probability density function*, which shows the

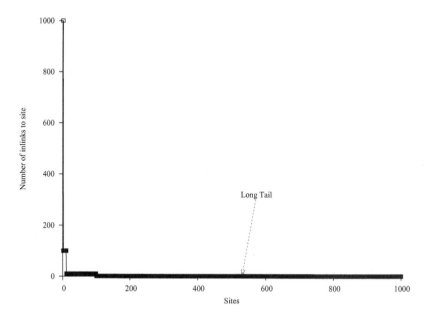

Figure 10.1 Indegree–rank plot of demonstration data, with Long Tail shown

---

[2]Other terminology that is used to describe this phenomenon is 'the 80/20 rule' (20% of the items or products in the market gain 80% of sales) and 'winner takes all'.

probability of a random draw from the distribution (or random variable) taking a particular value in the distribution (if the distribution is discrete) or the probability of the random variable being within a particular range of values (if the distribution is continuous). Figure 10.2 shows the Section 7.1.1 demonstration data plotted as a probability density function. The important thing to note is that now the units (websites) in the Long Tail are located on the left-hand side of the plot.

The probability density function in Figure 10.2 looks unusual because the data themselves are artificial. This is a heavily skewed distribution: most of the observations are grouped on the left-hand side (small-values of indegree), while a few observations are on the left-hand side with large values of indegree. This can be compared with the probability density function for the normal distribution, which is the familiar 'bell curve' (Figure 10.3). The distribution of any quantity (inlinks, sales) that exhibits a Long Tail looks very different from the normal (bell curve) distribution.

The important point here is that the term 'Long Tail' is used differently depending on whether you are looking at a rank plot or a probability density function. (We presented the indegree–rank plot in Figure 10.1, but if you were studying the frequency of words in the English language it would be a frequency–rank plot and if you were studying sales data it would be a sales-volume–rank plot.) In the former, as we have seen, the Long Tail refers to the many observations that have very small values for the variable of interest (e.g. the many websites with few or zero inlinks and the many products with

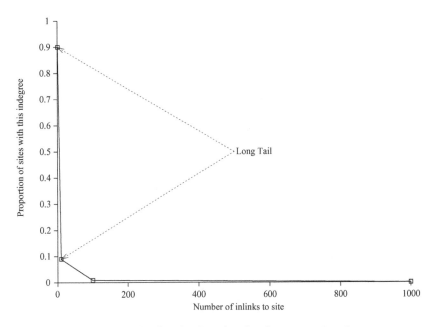

Figure 10.2 Probability density function for demonstration data

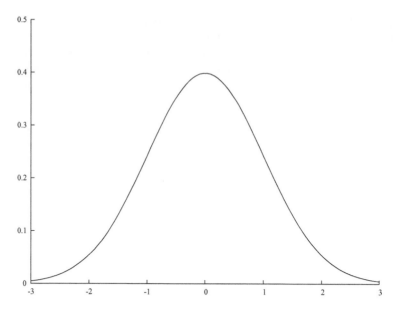

Figure 10.3 Probability density function – normal distribution

very small sales volume). In contrast, with the probability density function, the Long Tail is generally used to refer to the observations to the right-hand side of the distribution, that is, the few observations with very high values for the variable of interest (e.g. the 'superstar' websites that attract lots of inlinks, or blockbuster products that have very high sales volume). For the remainder of this section, we follow Anderson (2004, 2006) and use the term 'Long Tail' to refer to niche products with very small sales volumes.

An important question is: why might a Long Tail develop in a market? In Box 7.6 it was noted that power laws can emerge from feedback loops, self-organisation and behaviour in networks, and Box 7.2 presented the preferential attachment model that provides an explanation for why large-scale networks such as the web may exhibit power laws. But preferential attachment cannot be used to explain the emergence of Long Tails in a market (e.g. for books or music), and instead we can look to the 'economics of superstars' (Box 10.1) for insights into the emergence of superstars in markets.

10.1.2 Evidence for the Long Tail

The Long Tail proposition is that niche products can account for a much larger proportion of sales for an Internet business than they do for a traditional 'bricks-and-mortar' business. For example, it is possible for the online bookseller Amazon to have a much larger number of book titles on its

BOX 10.1   ECONOMICS OF SUPERSTARS

Why might a Long Tail emerge in a market? The economics of superstars – which originated in Rosen (1981) and Adler (1985) and was later elaborated by Frank and Cook (1995) – has been used to explain why in certain fields (e.g. the arts and sport) there is a concentration of output among a few individuals, and an associated marked skewness in the distribution of income, with very large rewards at the top. This explains why the probability density function for income of artists (for example) can be 'stretched out' in the right-hand tail, compared with the distribution of talent – that is, the differences in success (measured by income) can be far greater than the differences in talent. The economics of superstars can therefore be used to explain the emergence of highly unequal distributions of sales (and revenues) in a market.

Rosen (1981) proposes two factors that explain the emergence of superstars. First, consumer preferences lead to small differences in talent being magnified into larger earnings differences (with greater magnification at the top of the scale). This is because of imperfect substitution among different sellers (the example used by Rosen is if a surgeon is 10% more successful in saving lives, most people would pay more than a 10% premium for her services). Second, the existence of 'joint consumption technology' means that the cost of production does not rise in proportion to the size of a seller's market, allowing a concentration of output on few sellers who have most talent.

For Adler (1985), an interesting puzzle about stardom is that stars are often not more talented than many artists who are less successful, and he presents a model where large differences in earnings can exist with no differences in talent. Adler argues that consumers accumulate *consumption capital* (Stigler and Becker, 1977) in art, and the more consumption capital they possess, the greater the enjoyment from each encounter with the art. Thus, stardom is not due to the stars' superior talent but rather due to the need of consumers for a common culture (people are inherently social), that is, to consume the same art that other consumers do.

catalogue compared with a bricks–and–mortar bookseller, since the books can be stored cheaply in warehouses (instead of being displayed in shops), with customers purchasing the books via the Amazon website and having them delivered via mail.

A more radical part of the Long Tail argument is that the Internet is changing the distribution of sales such that the Long Tail is becoming 'fatter' or 'heavier', that is, the share of sales accounted for by niche products increases, with superstar products becoming less important. The commercial implication is that it will become profitable for companies to start targeting niches. However, the empirical evidence on this is not conclusive.

Brynjolfsson et al. (2009) present evidence that the Long Tail for Amazon became heavier between 2000 and 2008 (superstar products becoming less important), while Elberse (2008) – drawing on research by Elberse and Oberholzer-Gee (2008) – argues that while the tail is lengthening (more obscure products are being made available for purchase), it will remain extremely flat and the demand for blockbusters will continue to grow.

There are three ways of deriving a measure of whether the Long Tail of the sales distribution has changed (Brynjolfsson et al., 2010):

- *absolute Long Tail* – the number of titles in a particular category above an absolute cut-off (Brynjolfsson et al. (2003) used a cut-off of 100,000, which is the typical size of a large bricks-and-mortar bookstore);
- *relative Long Tail* – the relative share of sales above or below a certain rank (e.g. the percentage of sales accounted for by the top 10% of titles);
- *power law coefficient* – if the sales distribution data do follow a power law, then the regression coefficient for the log–log plot of the CCDF (the analogue for sales data of Figure 7.6) provides an indication of the heaviness of the tail of the sales distribution.

Brynjolfsson et al. (2010) note that if the number of titles has changed over time, then absolute and relative measures may provide different conclusions as to whether the tail of the sales distribution is getting heavier or lighter over time. For example, assume we are using a relative Long Tail measure of the percentage of sales accounted for by the top 10% of titles, and an absolute measure of the number of titles over a cut-off of 100. Assume a firm with 100 titles, with the top 20% accounting for 80% of sales, doubles the number of available titles but all the new titles have minimal sales. The relative Long Tail measure would indicate that the Long Tail had become less important, while the absolute measure would suggest the reverse.

10.2 INFLUENCE IN MARKETS

In Section 5.2 it was noted that there are three potential explanations for two socially connected people sharing the same behaviour (e.g. smoking). First, social influence is where one person's behaviour can be attributed to the influence of other people in their social network (e.g. friends, colleagues). Second, there may be social selection, with people becoming socially connected because they share particular preferences or characteristics (homophily) related to the behaviour of interest. Finally, there is the role of contextual factors, such as advertising that two people may both have seen.

In this section we revisit the topic of social influence, but we look at it from the point of view of product adoption (i.e. marketing). We also look at how contextual factors may influence the behaviour of a person, focusing in particular on *social* contextual factors. So this is where a person's decision is influenced by other peoples' behaviour, but they may be complete strangers (i.e. the influence of rating systems on behaviour).[3]

10.2.1 Referrals from friends

In marketing, there is a growing body of literature on the role of social influence in consumer decision-making. The term that is often used in this context is 'viral marketing'. This section reviews two examples of research into social influence in the context of online product adoption.

Studying social influence via a Facebook application

Aral and Walker (2010) use a Facebook application to study two aspects of online marketing: viral product design (the impact of different viral marketing design features on adoption behaviour) and statistical identification of social influence.

The authors identify two basic types of viral product design. *Personalised referrals* are when the Facebook application allows you to invite your friends via a personalised referral email to sign up for the application. These are referred to as 'active-personalised' designs since they involve more effort on behalf of the person making the referral (e.g. selecting which friends are going to receive the referral and what personalised message to include in the referral). *Automated broadcast notifications* are when the application automatically alerts your friends that you have started using the application. These are also known as 'passive-broadcast' designs because the person using the Facebook application need not actively make a decision as to which friends will receive the referral.

The theory suggests that active-personalised designs should be more effective per referral (the probability of a positive response – e.g. a friend then going on to install the Facebook application – for each referral made) than passive-broadcast designs because they allow for both targeting (you can select exactly which of your friends will receive an invite to join Facebook, and you are more likely to choose friends who you think would be interested in the service) and customisation (you can change the invitation message to suit the friend). Active-personalised viral marketing is thus leveraging the existence of homophily in social networks (Section 5.1). However, the fact that active-personalised designs involve effort on behalf of the person making the referral means that the greater effectiveness (in terms of product adoption) *per referral* might be offset by the fact that there are fewer active-personalised referrals being made.

It is therefore an empirical question as to which viral marketing design is most effective overall, and this is what Aral and Walker (2010) address. The

---

[3] Authors such as Salganik and Watts (2009) refer to this as 'social influence' but we use the term 'social contextual factors' here, to distinguish it from influence in social networks.

authors conduct a randomised field experiment (see Sections 2.1 and 5.2.1) by partnering with a company that develops a particular Facebook application called FLIXTER which allows users to share opinions about actors, movies, directors and the film industry in general.[4] Three experimental versions of FLIXTER were developed: the baseline control version and two versions containing active-personalised and passive-broadcast viral marketing features. When Facebook users signed up to FLIXTER they were randomly assigned one of the three versions of the application. The application also collected attribute data from the user's Facebook profiles, including data on their social networks and the attributes of their friends.

The randomised experiment ran for 44 days, and during this period 9,687 people signed up for the Facebook application, with 405 being assigned to the baseline control group, 4,600 being assigned to the passive-broadcast group and 4,682 being assigned to the active-personalised treatment group. The authors found that the number of peer adopters was 7 times higher (than baseline) for the passive-broadcast group and 10 times higher for the active-personalised group. Overall, they found (p. 19) that: 'Features that require more activity on the part of the user and are more personalized to recipients create greater marginal increases in the likelihood of adoption per message, but also generate fewer messages resulting in less total peer adoption in the network.'

Social influence in a social network site

Katona et al. (2011) use data from a central European social network site to identify the factors influencing the growth of its membership. The data were collected over a three-and-a-half-year period during which the only way a person could sign up for an account was to be first invited by an existing member. The authors argue that at the time of the study, word-of-mouth marketing (via receiving an invitation to join from an existing member) was the only way that people could find out about the service.

The authors know the date when each new member signed up; however, they do not know the source of the invitation(s) – they do not know which existing member(s) sent the invitation(s) that led to the application for membership. They make an assumption that for each member who joined, the online social network at the end of the period is an accurate representation of the social network that existed at the time he or she joined the

---

[4]As Aral and Walker (2010, p. 39) note, their randomised trial differs significantly from the standard approach. Generally, a randomised trial is used to quantify the effect of a changed condition on the behaviour of the person who has been treated. However, in this case the trial is being used to assess the impact of the treatment (particular viral marketing design) on the behaviour of the *friends* of the treated person (i.e. the impact on their tendency to adopt the product). Aral and Walker (2010) refer to this as an 'inside-out' randomised design.

service.[5] Finally, the authors know for a given person who joined the service, which of their friends were themselves members of it, and the social network of those people as well.

The authors propose a model where the probability of adoption (signing up for membership) depends on three factors: ego network structure, characteristics of network neighbours who have already adopted ('influencer effect'), and characteristics of the individual who is choosing whether to adopt ('adopter effect').

Regarding ego network structure, they find that the probability of adoption increases with the number of neighbours who are adopters. They also find that the probability of adoption rises with increasing interconnectedness or clustering of neighbours who have adopted. This draws on network closure theory (Box 3.4) which posits that two actors who are both friends of a given person will have more influence over that individual if they themselves are also friends, compared with if they do not know each other.

Concerning the influencer effects, they show that adoption probability is negatively related to the average degree of neighbours who have already adopted. The reasoning behind this is that people have a constant amount of 'power' to influence others, and that this power is apportioned across all friends. Hence, the average intensity of friendship of an actor (and hence the power of the actor to influence the decisions of others) must decrease with the number of friends the actor has. They also found that younger people and females exert greater influence, with females exerting significantly higher influence among younger individuals.

Finally, regarding the adopter effects, both age and sex exert a significant (but small) effect on the probability of adoption. There is a positive correlation between age and the probability of adoption (this was counter-intuitive since it was expected that younger people would be more susceptible to influence).

10.2.2 Ratings systems

Social contextual factors such as the views and opinions of the general public are important drivers of individual choice. For example, it has long been understood by marketers that rankings such as the Billboard Charts exert influence on the music-buying public. But prior to the Internet it was very difficult to study these types of large-group or collective social dynamics. This section looks at two studies that aim to provide further understanding of the role of the general public in influencing individual choice.

Artificial cultural markets

Salganik and Watts (2009) focus on the 'puzzling nature' of success or failure in cultural markets – books, music, TV shows, movies. They are

---

[5]The authors provide robustness checks of the above assumption, and several other assumptions that are key to their empirical approach.

interested in two stylised facts of cultural markets (both of these are relevant to the economics of superstars – Box 10.1). First, success is very unequal (see the 'superstar' phenomenon discussed in Section 10.1.1 and the related 'winner-takes-all' discussion in Section 9.2.2). Second, cultural markets are very unpredictable: it is hard to predict 'hits', and the authors cite the fact that eight publishers rejected the first book in what became the hugely successful *Harry Potter* series, suggesting that superstars may not have higher quality than the rest.

The paper explores the possibility that preferences are influenced by social processes that we term here 'social contextual factors' and the authors term 'collective social dynamics'. In particular, they are interested in establishing the role of cumulative advantage – the fact that a superstar book/movie/TV standard show/singer may emerge not because of quality difference but because there is some initial random event (luck) that gets some people interested and this snowballs into success.

They present a 'thought piece' on being able to 'rerun history' with regard to the *Harry Potter* book series to test whether social processes were the main factor in its success rather than the attributes of the books. Of course such a rerun of history is not possible, but the authors instead create an artificial cultural 'market' – a website where participants listen to, rate and download 48 songs by unknown bands.

They implement a 'multiple-worlds' experimental design, with 2,930 participants being randomly assigned into either a control 'world', where there is no information about the behaviour of others, or a number of treatment worlds where individuals are provided with information about the behaviour of those in their world (namely, the rating on a five-point scale of the various songs that are available for download), but not those in the other worlds. Participants can select and listen to a song in streaming mode, and then choose whether or not to download the song. The aim of the experiment was to see if ratings of songs influenced the download rate.

The authors find strong evidence of cumulative advantage: information about the popularity of a song affected the probability of a participant subsequently downloading the song (after having listened to it in streaming mode). They further note that while, at the margins, quality did affect download behaviour – truly good songs had higher rates of download regardless of whether social contextual information was provided, and vice versa for truly bad songs – for songs of average quality, social context had significant impact on download behaviour.

Comparing friend recommendations and general public ratings

Abbassi et al. (2011) conduct online experiments using Mechanical Turk (Box 2.1) to compare the effect of friend recommendations and ratings from the general public on individual behaviour.[6] They are able to assess the trade-off that people make between influence from friends (social influence) and influence from the general public (social contextual factors) by presenting Mechanical Turk workers with a hypothetical situation where they make a decision on whether to purchase a hotel booking online or watch a movie trailer, and they are provided with information on the general public rating of the product (number of stars) and also the number of friends recommending the product. They find that for a user deciding between two options, an additional rating star has a much greater influence on the probability of choosing the option, compared with an additional friend's recommendation.

10.2.3 Recommender systems

There are two basic types of recommender system. *Content-based systems* provide recommendations purely based on the attributes of the product (e.g. if you have purchased a novel you may be recommended other novels by the same author or of the same genre). *Collaborative filter-based* systems recommend products that similar customers bought (leading to Amazon's famous tagline: 'Customers who bought this also bought ...').

Parallelling research into whether the web has led to a fragmenting or cyberbalkanisation on the basis of political preferences (Section 7.3), there are opposing conjectures on the impact of recommendation systems on sales diversity, and hence whether the Long Tail of the distribution of sales is getting heavier or lighter (Section 10.1.2). The role of recommender systems in helping consumers find niche products could help to increase sales diversity, yet Mooney and Roy (2000) have suggested that recommender systems may reinforce the position of already-popular products, thus reducing sales diversity.

Fleder and Hosanagar (2007) use a formal model of recommender systems to show that some well-known systems can reduce sales diversity. For example, the authors show that since collaborative filters use sales data to provide recommendations, they cannot recommend products with limited historical sales data, and this could contribute to a 'rich-get-richer' or 'winner-takes-all' phenomenon (Section 9.2.2) for popular products, thus decreasing sales diversity.

10.3 CONCLUSION

This chapter has looked at Internet commerce and marketing. We introduced the concept of the Long Tail and showed how this is relevant to

---

[6]The authors refer to these as two components of social influence, but as noted above, in this book we refer to influence from the general public as 'social contextual factors'.

power law distributions (introduced in Section 7.1.1). While the Long Tail concept has been influential, the empirical evidence that businesses are shifting away from blockbusters and instead focusing on niche products is not conclusive and is the subject of ongoing research. The Long Tail concept is not just about the supply side (businesses changing their approach to selling products); the demand side or consumer behaviour is also highly relevant. In particular, the key argument of the Long Tail thesis is that sales diversity is increasing (consumers are purchasing more niche products, at the expense of blockbusters), and hence it is important to look at research into how the Internet may be changing consumer behaviour. This chapter reviewed research into recommender systems and sales diversity. It also looked at how Internet data are providing new insights into aspects of marketing: social influence (word-of-mouth marketing) and rating systems.

Further reading

For more on the Long Tail, see Anderson (2006).

References

Abbassi, Z., Aperjis, C., and Huberman, B. A. (2011). Swayed by friends or by the crowd? arXiv:1111.0307v2.

Ackland, R. (2005). Mapping the U.S. political blogosphere: Are conservative bloggers more prominent? Paper presented at BlogTalk Downunder, 20-21 May 2005, University of Technology Sydney. Available at: http://incsub.org/blogtalk/?page_id=49.

Ackland, R. (2009). Social network services as data sources and platforms for e-researching social networks. *Social Science Computer Review*, 27(4): 481–492.

Ackland, R. (2010a). 'Superstar' concentrations of scientific output and recognition. In W. H. Dutton, and P. W. Jeffreys (eds), *World Wide Research: Reshaping the Sciences and Humanities*. MIT Press, Cambridge, MA.

Ackland, R. (2010b). WWW hyperlink networks. In D. L. Hansen, B. Shneiderman, and M. A. Smith (eds), *Analyzing Social Media Networks with NodeXL: Insights from a Connected World*. Morgan-Kaufmann, Burlington, MA.

Ackland, R. and Evans, A. (2005). The visibility of abortion-related information on the World Wide Web. Presented to the Australian Sociological Association Conference, University of Tasmania, Sandy Bay Campus, 5–8 December.

Ackland, R. and O'Neil, M. (2011). Online collective identity: The case of the environmental movement. *Social Networks*, 33: 177–190.

Ackland, R. and Shorish, J. (2009). Network formation in the political blogosphere: An application of agent-based simulation and e-research tools. *Computational Economics*, 34(4): 383–398.

Ackland, R., Gibson, R., Lusoli, W., and Ward, S. (2010). Engaging with the public? Assessing the online presence and communication practices of the nanotechnology industry. *Social Science Computer Review*, 28(4): 443–465.

Adamic, L. (1999). The small world web. In *Proceedings of the 3rd European Conference on Digital Libraries,* Lecture Notes in Computer Science 1696, pp. 443–452 Springer, New York.

Adamic, L. (2002). Zipf, Power-laws, and Pareto -a ranking tutorial. Online tutorial available at: http://www.hpl.hp.com/research/idl/papers/ranking/ranking.html.

Adamic, L. and Glance, N. (2005). The political blogosphere and the 2004 U.S. election: Divided they blog. In *Proceedings of the 3rd International Workshop on Link Discovery (LINKDD 2005)*, pp. 6–43. Available at: http://doi.acm.org/10.1145/1134271.1134277.

Adamic, L. and Huberman, B. (2000). Power-law distribution of the World Wide Web. *Science*, 287: 2115.

Adler, M. (1985). Stardom and talent. *American Economic Review*, 75(1): 208–212.

Alford, R., Funk, C., and Hibbing, J. (2005). Are political orientations genetically transmitted? *American Political Science Review*, 99: 153–167.

Almind, T. and Ingwersen, P. (1997). Informetric analyses on the World Wide Web: Methodological approaches to 'webometrics'. *Journal of Documentation*, 55(5): 404–426.

Anderson, C. (2004). The Long Tail. *Wired*, October. Available at: http://www.wired.com/wired/archive/12.10/tail.html.

Anderson, C. (2006). *The Long Tail: Why the Future of Business Is Selling Less of More.* Hyperion, New York.

Aral, S. and Walker, D. (2010). Creating social contagion through viral product design: A randomized trial of peer influence in networks. New York University Stern School of Business Working Paper.

Aral, S., Muchnik, L., and Sundararajan, A. (2009). Distinguishing influence-based contagion from homophily-driven diffusion in dynamic networks. *Proceedings of the National Academy of Sciences*, 106(51): 21544–21549.

Babbie, E. R. (2007). *The Practice of Social Research,* 11th edition. Thomson Wadsworth, Belmont, CA.

Balicer, R. (2007). Modeling infectious diseases' dissemination through online roleplaying games: Virtual epidemiology. *Epidemiology*, 18(2): 260–261.

Barabási, A.-L. (2002). *Linked: The New Science of Networks.* Perseus Books Group, New York.

Barabási, A.-L. and Albert, R. (1999). Emergence of scaling in random networks. *Science*, 286: 509–512.

Barabási, A.-L., Albert, R., and Jeong, H. (2000). Scale-free characteristics of random networks: The topology of the World Wide Web. *Physica A*, 281: 69–77.

Barjak, F. and Thelwall, M. (2008). A statistical analysis of the web presences of European life sciences research teams. *Journal of the American Society for Information Science and Technology*, 59(4): 628–643.

Barnett, G.A. (2001). A longitudinal analysis of the international telecommunications network: 1978–1996. *American Behavioral Scientist*, 44(10): 1638–1655.

Barnett, G.A., Chung, C.J., and Park, H.W. (2011). Uncovering transnational hyperlink patterns and web-mediated contents: A new approach based on cracking .com domain. *Social Science Computer Review*, 29(3): 369–384.

Barok, D. (2011). Bitcoin: Censorship-resistant currency and domain name system to the people. Networked Media, Piet Zwart Institute, Rotterdam, July 20, 2011. Available at: http://pzwart3.wdka.hro.nl/mediawiki/images/6/64/Barok.bitcoin.pdf (accessed 11 July 2012).

Benford, R. D., Rochford, E. B., Snow, D. A., and Worden, S. K. (1986). Frame alignment processes, micromobilization, and movement participation. *American Sociological Review*, 51: 464–481.

Benkler, Y. (2006). *The Wealth of Networks: How Social Production Transforms Markets and Freedom.* Yale University Press, New Haven, CT.

Bennett, L.W. (2004). Communicating global activism. In W. van de Donk, B. Loader, P. G. Nixon, and D. Rucht (eds), *Cyberprotest: New Media, Citizens and Social Movements.* Routledge, London and New York.

Best, S. J. and Krueger, B. S. (2005). Analyzing the representativeness of Internet political participation. *Political Behavior*, 27: 183–216.

Best, S. J. and Krueger, B. S. (2009). Internet survey design. In N. Fielding, R. M. Lee, and G. Blank (eds), *Online Research Methods*. Sage, London.

Bianconi, G. and Barabási, A.-L. (2001a). Bose–Einstein condensation in complex networks. *Physical Review Letters*, 86: 5632–5635.

Bianconi, G. and Barabási, A.-L. (2001b). Competition and multiscaling in evolving networks. *Europhysics Letters*, 54: 436–442.

Bimber, B. (2001). Information and political engagement in America: The search for effects of information technology at the individual level. *Political Research Quarterly*, 54(1): 53–67.

Bimber, B., Flanagin, A. J., and Stohl, C. (2005). Reconceptualizing collective action in the contemporary media environment. *Communication Theory*, 15: 365–388.

Björneborn, L. and Ingwersen, P. (2004). Toward a basic framework for webometrics. *Journal of the American Society for Information Science and Technology*, 55(14): 1216–1227.

Borquez, F. and Ackland, R. (2012). Horses for courses? A comparative study of web crawlers for the social sciences. Mimeograph, The Australian National University.

Boulianne, S. (2009). Does internet use affect engagement? A meta-analysis of research. *Political Communication*, 26: 193–211.

Boxman, E., Graf, P. D., and Flap, H. (1991). The impact of social and human capital on the income attainment of Dutch managers. *Social Networks*, 13: 51–73.

Boyd, D. and Ellison, N. (2008). Social network sites: Definition, history, and scholarship. *Journal of Computer-Mediated Communication*, 13: 210–230.

Brent, E. (2009). Artificial intelligence and the internet. In N. Fielding, R. M. Lee, and G. Blank (eds), *Online Research Methods*. Sage, London.

Brock, W. A. and Durlauf, S. N. (2001). Discrete choice with social interactions. *Review of Economic Studies*, 68: 235–260.

Bruns, A. (2008). *Blogs, Wikipedia, Second Life, and Beyond: From Production to Produsage*. Peter Lang, New York.

Brynjolfsson, E., Hu, Y. J., and Smith, M. D. (2003). Consumer surplus in the digital economy: Estimating the value of increased product variety at online booksellers. *Management Science*, 49(11): 1580–1596.

Brynjolfsson, E., Hu, Y. J., and Smith, M. D. (2009). A longer tail? Estimating the shape of Amazon's sales distribution curve in 2008. Working Paper, MIT Sloan School of Management, Cambridge, MA.

Brynjolfsson, E., Hu, Y. J., and Smith, M. D. (2010). Research commentary – Long tails vs. superstars: The effect of information technology on product variety and sales concentration patterns. *Information Systems Research*, 21: 736–747.

Burt, R. (1992). *Structural Holes: The Social Structure of Competition*. Harvard University Press, Cambridge, MA.

Burt, R. (2011). Structural holes in virtual worlds. Booth School of Business, University of Chicago, Working Paper.

Burton, K., Java, A., and Soboroff, I. (2009). The ICWSM 2009 Spinn3r dataset. In *Proceedings of the Third Annual Conference on Weblogs and Social Media (ICWSM 2009)*.

Caldas, A., Schroeder, R., Mesch, G., and Dutton, W. (2008). Patterns of information search and access on the World Wide Web: Democratizing expertise or creating new hierarchies? *Journal of Computer-Mediated Communication*, 13: 769–793.

Casilli, A. A. and Tubaro, P. (2011). Why net censorship in times of political unrest results in more violent uprisings: A social simulation experiment on the UK riots. Working paper. Available at: http://ssrn.com/abstract=1909467.

Castells, M. (1996). *The Rise of the Network Society. The Information Age: Economy, Society and Culture, Vol. I.* Blackwell, London.

Castells, M. (1997). *The Power of Identity. The Information Age: Economy, Society and Culture, Vol. II.* Blackwell, Oxford.

Castells, M. (2004). *The Power of Identity. The Information Age: Economy, Society and Culture, Vol. II,* 2nd edition. Blackwell, Oxford.

Castronova, E. (2007). *Exodus to the Virtual World: How Online Games Will Change Reality.* Palgrave Macmillan, New York.

Castronova, E. (2009). *Synthetic Worlds: The Business and Culture of Online Games.* University of Chicago Press, Chicago.

Castronova, E., Williams, D., Shen, C., Ratan, R., Xiong, L., Huang, Y., and Keegan, B. (2009). As real as real? Macroeconomic behavior in a large-scale virtual world. *New Media & Society*, 11(5): 685–707.

Centola, D. (2010). The spread of behavior in an online social network experiment. *Science*, 329: 1194–1197.

Centola, D. (2011). An experimental study of homophily in the adoption of health behavior. *Science*, 334: 1269–1272.

Centola, D. and Macy, M. W. (2007). Complex contagion and the weakness of long ties. *American Journal of Sociology*, 113(3): 702–734.

Cha, M., Haddadi, H., Benevenuto, F., and Gummad, K. P. (2010). Measuring user influence on Twitter: The million follower fallacy. Fourth International AAAI Conference on Weblogs and Social Media, Washington, DC.

Chen, P. (2000). Pornography, protection, prevarication: The politics of internet censorship. *University of New South Wales Law Journal*, 23(1). Available at: http://www.austlii.edu.au/au/journals/UNSWLJ/2000/4.html (accessed 12 June 2012).

Christakis, N. A. and Fowler, J. H. (2007). The spread of obesity in a large social network over 32 years. *New England Journal of Medicine*, 357: 370–379.

Christakis, N. A. and Fowler, J. H. (2009). *Connected: The Surprising Power of Our Social Networks and How They Shape Our Lives.* Little, Brown, New York.

Clarke, R. (2004). Origins and nature of the Internet in Australia. Available at http://www.anu.edu.au/people/Roger.Clarke/II/OzI04.html.

Clauset, A. (2011). Inference, models and simulation for complex systems – lecture notes. Available at: http://tuvalu.santafe.edu/_aaronc/courses/7000/csci7000-001_2011_L2.pdf (accessed 4 September 2012).

Cohen, S. (2004). Social relationships and health. *American Psychologist*, 59(8): 676–684.

Cohen-Cole, E. and Fletcher, J. M. (2008). Is obesity contagious? Social networks vs. environmental factors in the obesity epidemic. *Journal of Health Economics*, 27: 1382–1387.

Coleman, J. (1958). Relational analysis: The study of social organizations with survey methods. *Human Organization*, 17: 28–36.

Costello, E., Compton, S., Keeler, G., and Angold, A. (2003). Relationship between poverty and psychopathology: A natural experiment. *Journal of the American Medical Association*, 290(15): 2023–2029.

Currarini, S., Jackson, M. O., and Pin, P. (2009). An economic model of friendship: Homophily, minorities and segregation. *American Economic Review*, 77(4): 1003–1045.

Dahlman, C. J. (1979). The problem of externality. *Journal of Law and Economics*, 21(2): 141–162.

Davenport, E. and Cronin, B. (2000). The citation network as a prototype for representing trust in virtual environments. In B. Cronin and H. Atkins (eds), *The Web of Knowledge: A Festschrift in Honor of Eugene Garfield*. Information Today, Metford, NJ.

Deibert, R. J., Palfrey, J. G., Rohozinski, R., and Zittrain, J. (2008). *Access Denied: The Practice and Policy of Global Internet Filtering*. MIT Press, Cambridge, MA.

della Porta, D. and Diani, M. (2006). *Social Movements: An Introduction*. Blackwell, Oxford.

Demil, B. and Lecoq, X. (2006). Neither market nor hierarchy nor network: The emergence of bazaar governance. *Organization Studies*, 27(10): 1447–1466.

Depken, C. A. (2005). The demand for censorship. *First Monday*, 11(9). Available at http://www.firstmonday.dk/issues/issue11_9/depken/index.html.

Diani, M. (1992). The concept of social movement. *Sociological Review*, 40(1): 1–25.

Diani, M. and Bison, I. (2004). Organizations, coalitions, and movements. *Theory and Society*, 33: 281–309.

Diani, M. and McAdam, D. (2003). *Social Movements and Networks: Relational Approaches to Collective Action*. Oxford University Press, Oxford.

DiMaggio, P., Hargittai, E., Neuman, W., and Robinson, J. (2001). Social implications of the internet. *Annual Review of Sociology*, 27: 307–336.

Dodge, M. and Kitchin, R. (2000). *Mapping Cyberspace*. Routledge, London.

Durkheim, E. (1964). *The Division of Labor in Society*. Free Press, New York.

Dutton, W. H. (2005). The Internet and social transformation: Reconfiguring access. In W. H. Dutton, B. Kahin, R. O'Callaghan, and A. W. Wyckoff (eds), *Transforming Enterprise*. MIT Press, Cambridge, MA.

Dutton, W. H. and Jeffreys, P. W. (2010). *World Wide Research: Reshaping the Sciences and Humanities*. MIT Press, Cambridge, MA.

Easley, D. and Kleinberg, J. (2010). *Networks, Crowds, and Markets: Reasoning about a Highly Connected World*. Cambridge University Press, Cambridge.

Elberse, A. (2008). Should you invest in the long tail? *Harvard Business Review*, 86(7/8): 88–96.

Elberse, A. and Oberholzer-Gee, F. (2008). Superstars and underdogs: An examination of the long-tail phenomenon in video sales. Unpublished manuscript.

Elliott, H. (1997). The use of diaries in sociological research on health experience. *Sociological Research Online*, 2(2). Available at http://www.socresonline.org.uk/socresonline/2/2/7.html (accessed May 2012).

Enyon, R., Fry, J., and Schroeder, R. (2009). The ethics of Internet research. In N. Fielding, R. M. Lee, and G. Blank (eds), *Online Research Methods*. Sage, London.

Escher, T., Margetts, H., Petricek, V., and Cox, I. (2006). Governing from the centre? Comparing the nodality of digital governments. Paper presented at the 2006 Annual Meeting of the American Political Science Association, Chicago, 31 August–4 September.

Ess, C. (2002). Ethical decision-making and Internet research: Recommendations from the AoIR ethics working committee. Association of Internet Researchers (AoIR). Available at http://aoir.org.reports/ethics.pdf (accessed 26 March 2013).

Evans, J. H. (1997). Multi-organizational fields and social movement organization frame content: The religious pro-choice movement. *Sociological Inquiry*, 67(4): 451–469.

Faas, T. and Schoen, H. (2006). Putting a questionnaire on the web is not enough – a comparison of online and offline surveys conducted in the context of the German federal election 2002. *Journal of Official Statistics*, 22(2): 177–190.

Faust, K. and Skvoretz, J. (2002). Comparing networks across space and time, size and species. *Sociological Methodology*, 32: 267–299.

Fielding, N. G., Lee, R. M., and Blank, G. (2008). *Sage Handbook of Internet and Online Research Methods*. Sage, London.

Fiore, A. and Donath, J. (2005). Homophily in online dating: When do you like someone like yourself? CHI 2005, 2–7 April, Portland, OR.

Fischer, C. S. (1997). Technology and community: Historical complexities. *Sociological Inquiry*, 67(1): 113–118.

Fischer, C. S. (2009). The 2004 GSS finding of shrunken social networks: An artifact? *American Sociological Review*, 74(4): 657–669.

Fleder, D. and Hosanagar, K. (2007). Blockbuster culture's next rise or fall: The impact of recommender systems on sales diversity. *Management Science*, 55(5): 697–712.

Flew, T. (2007). A citizen journalism primer. In Proceedings Communications Policy Research Forum 2007, University of Technology, Sydney. Available at: http://eprints.qut.edu.au/10232/01/10232.pdf (accessed 6 September 2012).

Flew, T. (2008). *New Media: An Introduction*. Oxford University Press, Oxford.

Foot, K. and Schneider, S. (2004). The web as an object of study. *New Media & Society*, 6(1): 114–122.

Foot, K. A., Schneider, S. M., Dougherty, M., Xenos, M., and Larsen, E. (2003). Analyzing linking practices: Candidate sites in the 2002 U.S. electoral web sphere. *Journal of Computer-Mediated Communication*, 8(4). Available at http://jcmc.indiana.edu/vol8/issue4/foot.html (accessed 22 December 2006).

Frank, O. and Strauss, D. (1986). Markov graphs. *Journal of the American Statistical Association*, 81(395): 832–842.

Frank, R. H. and Cook, P. J. (1995). *The Winner-Take-All Society*. Free Press, New York.

Fricker, R. D. (2009). Sampling methods for web and e-mail surveys. In N. Fielding, R. M. Lee, and G. Blank (eds), *Online Research Methods*. Sage, London.

Fukuyama, F. (1999a). *The Great Disruption: Human Nature and the Reconstitution of Social Order*. Free Press, New York.

Fukuyama, F. (1999b). Social capital and civil society. Delivered at the International Monetary Fund Conference on Second Generation Reforms, 8–9 November. Available at http://www.imf.org/external/pubs/ft/seminar/1999/reforms/fukuyama.htm. Accessed: June 2012.

Fulk, J., Flanagin, A., Kalman, M., Monge, P., and Ryan, T. (1996). Connective and communal public goods in interactive communication systems. *Communication Theory*, 6: 60–87.

Gaiser, T. J. (2009). Online focus groups. In N. Fielding, R. M. Lee, and G. Blank (eds), *Online Research Methods*. Sage, London.

Garrett, R. K. (2006). Protest in an information society. A review of literature on social movements and new ICTs. *Information, Communication & Society*, 9(2): 202–224.

Gibson, R., Margolis, M., Resnick, D., and Ward, S. (2003). Election campaigning on the WWW in the US and UK: A comparative analysis. *Party Politics*, 9(1): 47–76.

Gladwell, M. (2010). Small change: Why the revolution will not be tweeted. *The New Yorker* 4 October, pp. 42–49.

Gladwell, M. (2011). Does Egypt need Twitter? *The New Yorker*, 2 February. Available at:http://www.newyorker.com/online/blogs/newsdesk/2011/02/does-egypt-need-twitter.html.

Goffman, E. (1974). *Frame Analysis*. Harvard University Press, Cambridge, MA.

Gonzalez-Bailon, S. (2009). Opening the black box of link formation: Social factors underlying the structure of the web. *Social Networks*, 31: 271–280.

Goode, L. (2009). Social news, citizen journalism and democracy. *New Media & Society*, 11(8): 1287–1305.

Govcom.org (1995). IssueCrawler.net instructions of use. Available at: http://www.govcom.org/Issuecrawler_instructions.htm.

Granovetter, M. (1973). The strength of weak ties. *American Journal of Sociology*, 78(6): 1360–1380.

Granovetter, M. (1985). Economic action and social structure: The problem of embeddedness. *American Journal of Sociology*, 91(3): 481–510.

Griscom, R. (2002). Why are online personals so hot? *Wired* 10.11. Available at: http://www.wired.com/wired/archive/10.11/view_pr.html (accessed 1 September 2012).

Gusfield, J. R. (1975). *Community: A Critical Response*. Harper, New York.

Habermas, J. (1989). *The Structural Transformation of the Public Sphere: An Inquiry into a Category of Bourgeois Society*. MIT Press, Cambridge, MA. English translation (by Thomas Burger) of 1962 book.

Hampton, K., Sessions, L., and Her, E. J. (2011). Core networks, social isolation, and new media: Internet and mobile phone use, network size, and diversity. *Information, Communication & Society*, 14(1): 130–155.

Hanneman, R. A. and Riddle, M. (2005). Introduction to social network methods. University of California, Riverside. Published in digital form at http://faculty.ucr.edu/~hanneman.

Hansen, D. L., Shneiderman, B., and Smith, M. A. (2010a). *Analyzing Social Media Networks with NodeXL: Insights from a Connected World*. Morgan-Kaufmann, Burlington, MA.

Hansen, D. L., Shneiderman, B., and Smith, M. A. (2010b). Thread networks: Mapping message boards and email lists. In D. L. Hansen, B. Shneiderman, and M. A. Smith (eds), *Analyzing Social Media Networks with NodeXL: Insights from a Connected World*. Morgan-Kaufmann, Burlington, MA.

Hargittai, E., Gallo, J., and Kane, M. (2008). Cross-ideological discussions among conservative and liberal bloggers. *Public Choice*, 134: 67–86.

Heider, F. (1958). *The Psychology of Interpersonal Relations*. Wiley, New York.

Herring, S. (1996). Linguistic and critical analysis of computer-mediated communication: Some ethical and scholarly considerations. *Information Society*, 12(2): 153–168.

Herring, S. (2010). Web content analysis: Expanding the paradigm. In J. Hunsinger, M. Allen, and L. Klastrup (eds), *The International Handbook of Internet Research*. Springer, Berlin.

Hewson, C., Yule, P., Laurent, D., and Vogel, C. (2003). *Internet Research Methods: A Practical Guide for the Behavioural and Social Sciences*. Sage, London.

Hindman, M. (2009). *The Myth of Digital Democracy*. Princeton University Press, Princeton, NJ.

Hindman, M., Tsioutsiouliklis, K., and Johnson, J. (2003). 'Googlearchy': How a few heavily-linked sites dominate politics on the Web. Mimeograph, Princeton University, 2003.

Hinduja, S. and Patchin, J. W. (2008). Personal information of adolescents on the internet: A quantitative content analysis of MySpace. *Journal of Adolescence*, 31(1): 125–146.

Hine, C. (2009). Virtual ethnography: Modes, varieties, affordances. In N. Fielding, R. M. Lee, and G. Blank (eds), *Online Research Methods*. Sage, London.

Hitsch, G., Hortacsu, A., and Ariely, D. (2005). What makes you click: An empirical analysis of online dating. Mimeograph, University of Chicago.

Hochschild, J. (1997). *The Time Bind: When Work Becomes Home and Home Becomes Work*. Henry Holt, New York.

Hoffman, A. J. and Bertels, S. (2007). Organizational sets, populations and fields: Evolving board interlocks and environmental NGOs. Ross School of Business Working Paper Series, Working Paper No. 1074.

Hogan, B. (2010). Visualizing and interpreting Facebook networks. In D. L. Hansen, B. Shneiderman, and M. A. Smith (eds), *Analyzing Social Media Networks with NodeXL: Insights from a Connected World*. Morgan-Kaufmann, Burlington, MA.

Hood, C. (1983). *The Tools of Government*. Macmillan, London.

Hood, C. and Margetts, H. (2007). *The Tools of Government in the Digital Age*. Macmillan, London.

Hookway, N. (2008). 'Entering the blogosphere': Some strategies for using blogs in social research. *Qualitative Research*, 8: 91–113.

Howard, P. N., Duffy, A., Freelon, D., Hussain, M., Mari, W., and Mazaid, M. (2011). Opening closed regimes: What was the role of social media during the Arab

Spring? Project on Information Technology and Political Islam, Research Memo 2011.1. University of Washington, Seattle. Available at: http://pitpi.org/index. php/2011/09/11/ opening-closed-regimes-what-was-the-role-of-social-media-during-the-arab-spring/ (accessed 4 September 2012).

Hunt, S. A. and Benford, R. D. (2004). Collective identity, solidarity and commitment. In D. Snow, S. Soule, and H. Kriesi (eds), *The Blackwell Companion to Social Movements*. Blackwell, London and New York.

Jackman, S. (2005). Pooling the polls over an election campaign. *Australian Journal of Political Science*, 40(4): 499–517.

Jackson, M. H. (1997). Assessing the structure of communication on the World Wide Web. *Journal of Computer-Mediated Communication*, 3(1): 273–299. Available at http://jcmc.indiana.edu/vol3/issue1/jackson.html.

Janetsko, D. (2009). Nonreactive data collection on the Internet. In N. Fielding, R. M. Lee, and G. Blank (eds), *Online Research Methods*. Sage, London.

Johnson, J. and Bytheway, B. (2001). An evaluation of the use of diaries in a study of medication in later life. *International Journal of Social Research Methodology*, 4(3): 183–204.

Kaczmirek, L. (2009). Internet survey software tools. In N. Fielding, R. M. Lee, and G. Blank (eds), *Online Research Methods*. Sage, London.

Katona, Z., Zubcsek, P., and Sarvary, M. (2011). Network effects and personal influences: Diffusion of an online social network. *Journal of Marketing Research*, 48(3): 425–443.

Katz, L., Kling, J., and Liebman, J. (2001). Moving to opportunity in Boston: Early results of a randomized mobility experiment. *Quarterly Journal of Economics*, 116: 607–654.

Katz, N., Lazer, D., Arrow, H., and Contractor, N. (2004). Network theory and small groups. *Small Group Research*, 35(3): 307–332.

Kleinberg, J. (1999). Authoritative sources in a hyperlinked environment. *Journal of the ACM*, 46(5): 604–632.

Kocher, R. (2000). Representative survey on Internet content concerns in Australia, Germany and the United States of America. In J. Walterman and M. Machill (eds), *Protecting Our Children on the Internet: Towards a New Culture of Responsibility*. Bertelsmann Foundation: Gutersloh.

Krackhardt, D. (1994). Graph theoretical dimensions of informal organizations. In M. J. Prietula (ed.), *Computational Organization Theory*. Lawrence Erlbaum Associates, Hillsdale NJ.

Kraut, R., Patterson, M., Lundmark, V., Kiesler, S., Mukhopadhyay, T., and Scherlis, W. (1998). Internet paradox: A social technology that reduces social involvement and psychological well-being? *American Psychologist*, 53(9): 1017–1031.

Kraut, R., Kiesler, S., Boneva, B., Cummings, J., Helgeson, V., and Crawford, A. (2002). Internet paradox revisited. *Journal of Social Issues*, 58(1): 49–74.

Krippendorff, K. (2004). *Content Analysis: An Introduction to Its Methodology*. Sage, Thousand Oaks, CA.

Kroh, M. and Neiss, H. (2009). Internet access and political engagement: Self-selection or causal effect? APSA 2009 Toronto Meeting Paper. Available at: http://ssrn.com/abstract=1451368 (accessed 20 July 2012).

Kwak, H., Lee, C., Park, H., and Moon, S. (2010). What is Twitter, a social network or a news media? In *Proceedings of the 19th International Conference on World Wide Web*, pp. 591–600, ACM Press, New York.

Lambe, J. L. (2004). Who wants to censor pornography and hate speech? *Mass Communication & Society*, 7(3): 279–299.

Lampe, C., Ellison, N., and Steinfield, C. (2006). A Face(book) in the crowd: Social searching vs. social browsing. In *Proceedings of CSCW-2006*. ACM Press, New York.

Latour, B. (2005). *Reassembling the Social: An Introduction to Actor-Network Theory*. Oxford University Press, Oxford.

Laumann, E., Marsden, P. V., and Prensky, D. (1983). The boundary specification problem in network analysis. In R. S. Burt and M. J. Minor (eds), *Applied Network Analysis*. Sage Publications, London.

Lazarsfeld, P. F. and Merton, R. K. (1954). Friendship as a social process: A substantive and methodological analysis. In M. Berger (ed.), *Freedom and Control in Modern Society*, pp. 18–66. Van Nostrand, New York.

Lazer, D., Rubineau, B., Katz, N., Chetkovich, C., and Neblo, M. (2008). Networks and political attitudes: Structure, influence, and co-evolution. Working Paper, Harvard University.

Lazer, D., Pentland, A., Adamic, L., Aral, S., Barabàsi, A.-L., Brewer, D., Christakis, N., Contractor, N., Fowler, J., Gutmann, M., Jebara, T., King, G., Macy, M., Roy, D., and Van Alstyne, M. (2009). Computational social science. *Science*, 323: 721–723.

Lee, S., Monge, P., Bar, F., and Matei, S. A. (2007). The emergence of clusters in the global telecommunications network. *Journal of Communication*, 57(3): 415–434.

Lenhart, A. and Madden, M. (2007). Teens, privacy, and online social networks. Pew Internet and American Life Project Report. Available at http://www.pewinternet.org/pdfs/PIP_Teens_Privacy_SNS_Report_Final.pdf (accessed 30 July 2007).

Lerner, J. and Tirole, J. (2002). Some simple economics of open source. *Journal of Industrial Economics*, 50(2): 197–234.

Lewis, K., Kaufman, J., Gonzalez, M. J., Wimmer, A., and Christakis, N. A. (2008). Tastes, ties, and time: A new social network dataset using Facebook.com. *Social Networks*, 30: 330–342.

Lin, N. (1999). Building a network theory of social capital. *Connections*, 22(1): 28–51.

Lin, N. (2001). *Social Capital: A Theory of Social Structure and Action*. Cambridge University Press, Cambridge.

Lofgren, E. and Fefferman, N. (2007). The untapped potential of virtual game worlds to shed light on real world epidemics. *Lancet, Infectious Diseases*, 7(9): 625–629.

Lofland, L. H. (1973). *A World of Strangers: Order and Action in Urban Public Space*. Basic Books, New York.

Lusher, D. and Ackland, R. (2011). A relational hyperlink analysis of an online social movement. *Journal of Social Structure*, 12(5). Available at http://www.cmu.edu/joss/content/articles/volume12/Lusher.

Madden, M. and Lenhart, A. (2006). Online dating. Pew Internet & American Life Project, 5 March, Available at http://pewinternet.org/Reports/2006/Online-Dating.aspx (accessed 4 September 2012).

Manfreda, K. L., Bosnjak, M., Berzelak, J., Haas, I., and Vehovar, V. (2008). Web surveys versus other survey modes – a meta-analysis comparing response rates. *International Journal of Market Research*, 50(1): 79–104.

Manski, C. F. (1993). Identification of endogenous social effects: The reflection problem. *Review of Economic Studies*, 60: 531–542.

Margetts, H. (2009). The internet and public policy. *Policy & Internet*, 1(1): 1–21.

Margolis, M., Resnick, D., and Wolfe, J. (1999). Party competition on the Internet: Minor versus major parties in the UK and USA. *Harvard International Journal of Press/Politics*, 4(4): 24–47.

Markham, A. and Buchanan, E. (2012). Ethical decision-making and Internet research: Recommendations from the AOIR ethics working committee (Version 2). Association of Internet Researchers (AOIR). Available at http://aoir.org/reports/ethics2.pdf (accessed 26 March 2013).

Matzat, U. (2004a). Academic communication and Internet discussion groups: Transfer of information or creation of social contacts? *Social Networks*, 26: 221–255.

Matzat, U. (2004b). Cooperation and community on the Internet: Past issues and present perspectives for theoretical-empirical internet research. *Analyse & Kritik*, 26: 63–90.

Mayer, A. and Puller, S. L. (2008). The old boy (and girl) network: Social network formation on university campuses. *Journal of Public Economics*, 92: 329–347.

McCarthy, J. and Zald, M. N. (2002). The enduring vitality of the resource mobilization theory of social movements. In J. Turner (ed.), *Handbook of Sociological Theory*. Kluwer Academic/Plenum Publishers, New York.

McKee, R., Mutrie, N., Crawford, F., and Green, B. (2007). Promoting walking to school: Results of a quasi-experimental trial. *Journal of Epidemiology and Community Health*, 61(9): 818–823.

McNutt, K. (2006). Research note: Do virtual policy networks matter? Tracing network structure online. *Canadian Journal of Political Science/Revue Canadienne de Science Politique*, 39(2): 391–405.

McPherson, M., Smith-Lovin, M., and Cook, J. M. (2001). Birds of a feather: Homophily in social networks. *Annual Review of Sociology*, 27: 415–444.

McPherson, M., Smith-Lovin, M., and Brashears, M. (2006). Social isolation in America. *American Sociological Review*, 71(3): 353–375.

McPherson, M., Smith-Lovin, M., and Brashears, M. (2009). Models and marginals. *American Sociological Review*, 74(4): 670–681.

Melucci, A. (1995). The process of collective identity. In H. Johnston and B. Klandemans (eds), *Social Movements and Culture*. University of Minnesota Press, Minneapolis.

Mercken, L., Snijders, T., Steglich, C., Vartiainen, E., and de Vries, H. (2010). Dynamics of adolescent friendship networks and smoking behavior. *Social Networks*, 32: 72–81.

Merton, R. K. (1968). The Matthew effect in science. *Science*, 159(3810): 56–63.

Monge, P. and Contractor, N. (2003). *Theories of Communication Networks*. Oxford University Press, New York.

Montgomery, J. D. (1991). Social networks and labor-market outcomes: Toward an economic analysis. *American Economic Review*, 81: 1408–1418.

Montgomery, J. D. (1992). Job search and network composition: Implications of the strength-of-weak-ties hypothesis. *American Sociological Review*, 57: 586–596.

Mooney, R. J. and Roy, L. (2000). Content-based book recommending using learning for text categorization. In *Proceedings of the 5th ACM Conference on Digital Libraries*, pp. 195–204, ACM Press, New York.

Mouw, T. (2006). Estimating the causal effect of social capital: A review of recent research. *Annual Review of Sociology*, 32: 79–102.

Neuendorf, K. A. (2002). *The Content Analysis Guidebook*. Sage, Thousand Oaks, CA.

Neuman, W. L. (2006). *Social Research Methods: Qualitative and Quantitative Approaches*, 6th edition. Allyn & Bacon, Boston.

Norris, P. (2001). *Digital Divide? Civic Engagement, Information Poverty and the Internet Worldwide*. Cambridge University Press, Cambridge.

Norris, P. (2005). The impact of the Internet on political activism: Evidence from Europe. *International Journal of Electronic Government Research*, 1: 20–39.

O'Connor, H., Madge, C., Shaw, R., and Wellens, J. (2009). Internet-based interviewing. In N. Fielding, R. M. Lee, and G. Blank (eds), *Online Research Methods*. Sage, London.

Olson, M. (1965). *The Logic of Collective Action*. Harvard University Press, Cambridge, MA.

O'Neil, M. (2009). *Cyberchiefs: Autonomy and Authority in Online Tribes*. Pluto Press, London.

Paolacci, G., Chandler, J., and Ipeirotis, P. (2010). Running experiments on Mechanical Turk. *Judgment and Decision Making*, 5(5): 411–419.

Park, H. W. (2003). Hyperlink network analysis: A new method for the study of social structure on the web. *Connections*, 25(1): 49–61.

Park, H. W., Kim, C. S., and Barnett, G. A. (2004). Socio-communicational structure among political actors on the web in South Korea: The dynamics of digital presence in cyberspace. *New Media & Society*, 6(3): 403–423.

Park, H. W., Barnett, G. A., and Chung, C. J. (2011). Structural changes in the 2003-2009 global hyperlink network. *Global Networks*, 11(4): 522–542.

Pattison, P. and Wasserman, S. (1999). Logit models and logistic regressions for social networks: II. Multivariate relations. *British Journal of Mathematical & Statistical Psychology*, 52: 169–193.

Pennock, D., Flake, G., Lawrence, S., Glover, E., and Giles, C. (2002). Winners don't take all: Characterizing the competition for links on the web. *Proceedings of the National Academy of Sciences*, 99(8): 5207–5211.

Pescosolido, B., Grauerholz, E., and Milkie, M. (1997). Culture and conflict: The portrayal of blacks in U.S. children's literature through the 20th century. *American Sociological Review*, 62: 443–464.

Petricek, V., Escher, T., Cox, I., and Margetts, H. (2006). The web structure of e-government – developing a methodology for quantitative evaluation. In *Proceedings of the 15th International Conference on World Wide Web*, ACM Press, New York.

Pfeil, U. and Zaphiris, P. (2010). Applying qualitative content analysis to study online support communities. *Universal Access in the Information Society*, 9(1): 1–16.

Pickerill, J. (2004). Rethinking political participation: Experiments in internet activism in Australia and Britain. In R. Gibson, A. Roemmele, and S. Ward (eds), *Electronic Democracy: Mobilisation, Organisation and Participation via New ICTs*. Routledge, London.

Portes, A. (1998). Social capital: Its origins and applications in modern sociology. *Annual Review of Sociology*, 22: 1–24.

Putnam, R. (1993). *Making Democracy Work: Civic Traditions in Modern Italy*. Princeton University Press, Princeton, NJ.

Putnam, R. (2000). *Bowling Alone: The Collapse and Revival of American Community*. Simon & Schuster, New York.

Reagans, R., Zuckerman, E., and McEvily, B. (2007). On firmer ground: The collaborative team as a strategic research site for verifying networked-based social-capital hypotheses. In J. E. Rauch (ed.), *The Missing Links: Formation and Decay of Economic Network*. Russell Sage Foundation, New York.

Reips, U.-D. (2002). Standards for Internet-based experimenting. *Experimental Psychology*, 49(4): 243–256.

Remler, D. K. and Van Ryzin, G. G. (2011). *Research Methods in Practice: Strategies for Description and Causation*. Sage Publications, London.

Resnick, P., Hansen, D., Riedl, J., Terveen, L., and Ackerman, M. (2005). Beyond threaded conversation. In *CHI'05 Extended Abstracts on Human Factors in Computing Systems* (Portland, OR, 2–7 April), pp. 2138–2139.

Rheingold, H. (1993). *The Virtual Community: Homesteading on the Electronic Frontier 2nd ediion*, 1st edition. Addison-Wesley, Reading, MA.

Rheingold, H. (2000). *The Virtual Community: Homesteading on the Electronic Frontier 2nd ediion*. MIT Press, London. Availlable at http://www.rheingold.com/vc/book/intro.

Riffe, D., Lacy, S., and Fico, F. (2005). *Analyzing Media Messages: Using Quantitative Content Analysis in Research*. Erlbaum, Mahwah, NJ.

Ritzer, G. and Jurgenson, N. (2010). Production, consumption, prosumption: The nature of capitalism in the age of the digital prosumer. *Journal of Consumer Culture*, 10(1): 13–36.

Robins, G., Pattison, P., and Wasserman, S. (1999). Logit models and logistic regressions for social networks: III. Valued relations. *Psychometrika*, 64(3): 371–394.

Robins, G., Pattison, P., Kalish, Y., and Lusher, D. (2007). An introduction to exponential random graph (p*) models for social networks. *Social Networks*, 29(2): 173–191.

Rogers, R. (2010a). Internet research: The question of method – a keynote address from the YouTube and the 2008 election cycle in the United States conference. *Journal of Information Technology & Politics*, 7: 241–260.

Rogers, R. (2010b). Mapping public web space with the IssueCrawler. In C. Brossard and B. Reber (eds), *Digital Cognitive Technologies: Epistemology and Knowledge Society*. Wiley, London.

Rogers, R. and Marres, N. (2000). Landscaping climate change: A mapping technique for understanding science and technology debates on the world wide web. *Public Understanding of Science*, 9(2): 141–163.

Rogers, R. and Zelman, A. (2002). Surfing for knowledge in the information society. In G. Elmer (ed.), *Critical Perspectives on the Internet*. Rowman & Littlefield, Lanham, MD.

Rosen, S. (1981). The economics of superstars. *American Economic Review*, 71(5): 845–858.

Ross, H. L. (1970). An experimental study of the negative income tax. *Child Welfare*, 49(10): 562–569.

Rucht, D. (2004). The quadruple 'A': Media strategies of protest movements since the 1960s. In W. van de Donk, B. Loader, P. G. Nixon, and D. Rucht (eds), *Cyberprotest: New Media, Citizens and Social Movements*. Routledge, London and New York.

Sacerdote, B. (2001). Peer effects with random assignment: Results for Dartmouth room-mates. *Quarterly Journal of Economics*, 116: 681–703.

Salganik, M. J. and Watts, D. J. (2009). Web-based experiments for the study of collective social dynamics in cultural markets. *Topics in Cognitive Science*, 1: 439–468.

Sanders, D., Clarke, H. D., Stewart, M. C., and Whiteley, P. (2007). Does mode matter for modeling political choice? Evidence from the 2005 British Election Study. *Political Analysis*, 15(3): 257–285.

Shadbolt, N., Berners-Lee, T., and Hall, W. (2006). The Semantic Web revisited. *IEEE Intelligent Systems*, 21(3): 96–101.

Shah, D. V., Kwak, N., and Holbert, R. L. (2001). 'Connecting' and 'disconnecting' with civic life: Patterns of Internet use and the production of social capital. *Political Communication*, 18: 141–162.

Shumate, M. and Dewitt, L. (2008). The North/South divide in NGO hyperlink networks. *Journal of Computer-Mediated Communication*, 13: 405–428.

Skvoretz, J. and Faust, K. (2002). Relations, species, and network structure. *Journal of Social Structure*, 3(3).

Smith, M. and Kollock, P. (1999). *Communities in Cyberspace*. Routledge, London.

Smith, M. A., Turner, T., and Gleave, E. (2007). Sharing social accounting metadata – lessons from NetScan. Paper presented at the Third International Conference on e-Social Science, 8–9 October, University of Michigan.

Snow, D. (2001). Collective identity and expressive forms. In N. J. Smelser and P. B. Baltes (eds), *International Encyclopedia of the Social and Behavioral Sciences*. Elsevier Science, London.

Soetevent, A. R. (2006). Empirics of the identification of social interactions: An evaluation of the approaches and their results. *Journal of Economic Surveys*, 20(2): 193–228.

Stigler, G. and Becker, G. (1977). De gustibus non est disuptandum. *American Economic Review*, 67: 76–90.

Sunstein, C. (2001). *Republic.com*. Princeton University Press, Princeton, NJ.

Tarrow, S. (2002). The new transnational contention: Organisations, coalitions, mechanisms. Presented to APSA Annual Meeting, Chicago.

Taylor, M. (1982). *Community, Anarchy and Liberty*. Cambridge University Press, Cambridge.

Thelwall, M. (2004). *Link Analysis: An Information Science Approach*. Academic Press. Available at: http://linkanalysis.wlv.ac.uk.

Thelwall, M. (2006). Interpreting social science link analysis: A theoretical framework. *Journal of the American Society for Information Science and Technology*, 57(1): 60–68.

References

Thelwall, M. (2009a). Homophily in Myspace. *Journal of the American Society for Information Science and Technology*, 60(2): 219–231.

Thelwall, M. (2009b). *Introduction to Webometrics: Quantitative Research for the Social Sciences*. Morgan & Claypool, San Rafael, CA.

Thelwall, M. and Stuart, D. (2006). Web crawling ethics revisited: Cost, privacy and denial of service. *Journal of the American Society for Information Science and Technology*, 57(13): 1771–1779.

Thelwall, M., Vaughan, L., and Björneborn, L. (2005). Webometrics. *Annual Review of Information Science and Technology*, 39: 81–135.

Tolbert, C. J. and McNeal, R. S. (2003). Unraveling the effects of the Internet on political participation? *Political Research Quarterly*, 56(2): 175–185.

Turner, T., Smith, M., Fisher, D., and Welser, H. (2005). Picturing Usenet: Mapping computer-mediated collective action. *Journal of Computer-Mediated Communication*, 10(4). Available at http://jcmc.indiana.edu/vol10/issue4/turner.html.

Uzzi, B., Amaral, L., and Reed-Tsochas, F. (2007). Small-world networks and management science research: A review. *European Management Review*, 4: 77–91.

Van Alstyne, M. and Brynjolfsson, E. (2005). Global village or cyber-balkans? Modeling and measuring the integration of electronic communities. *Management Science*, 51(6): 851–868.

van de Donk, W., Loader, B., Nixon, P. G., and Rucht, D. (2004). Introduction: Social movements and ICTs. In W. van de Donk, B. Loader, P. G. Nixon, and D. Rucht (eds), *Cyberprotest: New Media, Citizens and Social Movements*. Routledge, London and New York.

Van den Bulte, C. and Lilien, G. (2001). Medical innovation revisited: Social contagion versus marketing effort. *American Journal of Sociology*, 106(5): 1409–1435.

van der Gaag, M. and Snijders, T. (2005). The resource generator: Measurement of individual social capital with concrete items. *Social Networks*, 27: 1–29.

Vehovar, V. and Manfreda, K. L. (2009). Overview: Online surveys. In N. Fielding, R. M. Lee, and G. Blank (eds), *Online Research Methods*. Sage, London.

Vergeer, M. and Hermans, L. (2008). Analysing online political discussions: Methodological considerations. *Javnost – The Public*, 15(2): 37–56.

Vogt, W. P., Gardner, D. C., and Haeffele, L. M. (2012). *When to Use What Research Design*. Guilford Press, New York.

Wallace, B. (2011). The rise and fall of Bitcoin. Available at: http://www.wired.com/magazine/2011/11/mf_bitcoin/all/1 (accessed 20 July 2012).

Wallerstein, I. (1974). *The Modern World System*. Academic Press, New York.

Wasserman, S. and Faust, K. (2004). *Social Network Analysis*. Cambridge University Press, Cambridge.

Wasserman, S. and Pattison, P. (1996). Logit models and logistic regressions for social networks: I. An introduction to Markov graphs and p*. *Psychometrika*, 61(3): 401–425.

Weber, M. (1922). *Wirtschaft und Gesellschaft*. JCB Mohr, Tubingen.

Wellman, B. (1998). Structural analysis: From method and metaphor to theory and substance. In B. Wellman and S. Berkowitz (eds), *Social Structures: A Network Approach*. Cambridge University Press, Cambridge.

Wellman, B. (2001). Physical place and cyberplace: The rise of personalized networking. *Journal of Urban and Regional Research*, 25(2): 227–252.

Wellman, B. and Wortley, S. (1990). Different strokes from different folks. *American Journal of Sociology*, 96(3): 558–588.

Welser, H., Gleave, E., Fisher, D., and Smith, M. (2007). Visualizing the signatures of social roles in online discussion groups. *Journal of Social Structure*, 8(2). Available at http://www.cmu.edu/joss/content/articles/volume8/Welser.

Whalen, R. (2011). The structure of federal egovernment: Using hyperlinks to analyze the .gov domain. Paper prepared for the 4th Annual Political Networks Conference, June 2011 University of Michigan, Ann Arbor. Available at: http://opensiuc.lib.siu.edu/pnconfs_2011/8.

Williams, D. (2007). The impact of time online: Social capital and cyberbalkanization. *Cyberpsychology & Behavior*, 10(3): 398–406.

Williams, D. (2010). The mapping principle, and a research framework for virtual worlds. *Communication Theory*, 20(4): 451–470.

Williams, D., Ducheneaut, N., Xiong, L., Zhang, Y., Yee, N., and Nickell, E. (2006). From tree house to barracks: The social life of guilds in World of Warcraft. *Games and Culture*, 1: 338–361.

Williams, D., Yee, N., and Caplin, S. E. (2008). Who plays, how much, and why? Debunking the stereotypical gamer profile. *Journal of Computer-Mediated Communication*, 13: 993–1018.

Wimmer, A. and Lewis, K. (2010). Beyond and below racial homophily: ERG models of a friendship network documented on Facebook. *American Journal of Sociology*, 116(2): 583–642.

Wu, S., Hofman, J., Mason, W. A., and Watts, D. J. (2011). Who says what to whom on Twitter. WWW 2011, 28 March–1 April, Hyderabad, India.

Yang, J. and Leskovec, J. (2011). Temporal variation in online media. In *ACM International Conference on Web Search and Data Mining (WSDM'11)*.

Yee, N., Bailenson, J., Urbanek, M., Chang, F., and Mergeta, D. (2007). The unbearable likeness of being digital: The persistence of nonverbal social norms in online virtual environments. *Cyberpsychology & Behavior*, 10(1): 115–121.

Zhao, S. (2006). Do internet users have more social ties? A call for differentiated analyses of internet use. *Journal of Computer-Mediated Communication*, 11(3).

Zhu, J. J. H., Mo, Q., Wang, F., and Lu, H. (2011). A random digit search (RDS) method for sampling of blogs and other user-generated content. *Social Science Computer Review*, 29(3): 327–339.

Zimmer, M. (2010). 'But the data is already public': On the ethics of research in Facebook. *Ethics & Information Technology*, 12(4): 313–325.

Zittrain, J. (2008). *The Future of the Internet – And How to Stop It.* Yale University Press, New Haven, CT.

Index

References in **bold** are to boxes, in *italics* are to figures, followed by a letter t are to tables and followed by a letter n are to footnotes.

80/20 rule, 164n, 168
  see also cumulative advantage

A-list bloggers, 130, 132
abortion debate, 130–1, *131*
academic output, 14, 156–7
accessibility, 140, 141
achievement, individual, 159–62
ACMA (Australian Communications and
    Media Authority), 144, 145
active-personalised referrals, 169–70
actor attributes, 55, 83
actor network theory, 83
actor-oriented models, 105
actor-relation effects, 59–60, 61, 84–5, 98
actors, 48
adjacency matrices, 52, 52t
advocacy groups, 81, 85
  see also NGOs
affiliation networks, 63, 65, 73
affirmation, social, 108, 143
agent-based modelling, 144
alters, **50**, 154
altruism, 10, 153
Amazon, 166–7, 168
anonymity
  of blogging, 38–9
  digital trace data, 44
  focus groups, 34
  Mechanical Turk, **42**
answer people, 153–4, *154*, *155*
AoIR (Association of Internet
    Researchers), 43
AOL (America Online), 6
Apache web server, **4**, 150
APIs (application programming
    interfaces), **69**, 72–3, 90, 91, 93
applied physics, xv, 14, 15, 120
Arab Spring, 142, 143
archives, 91, **92**, 132

Association of Internet Researchers
    (AoIR), 43
associational spaces, 83
assortative mixing, 97–8, **100**, 107, 132
asynchronous focus groups, 34
asynchronous online interviews, 31
attention
  flows of, 74–5
  political, 119–20, 121
attributes
  actor, 55, 83
  node, 49–50, 61, 141–2
  personal, 97–8, 101, 102, 103
Australia
  censorship, 144, 145
  elections, 29–30
  government websites, 141
Australian Communications and Media
    Authority (ACMA), 144, 145
authority
  government, 138, 142–6
  scholarly, 158–9
authorlines, 154, *154*, *155*
automated broadcast notifications,
    169–70
automated data collection tools, 46
  see also web crawlers
avatars, 7, **125**

baby boomers, 125
balance theory, 53
bazaar governance, 151, 151t, 152
behaviour, effects of web on,
    16–17, 143–4
betweenness centralisation, 77
betweenness centrality, 75–6
bias, sampling, 26, 28, 30, **42**, 145
bibliometrics, 14
Billboard Charts, 171
Bing, 90

Bitcoin, **139**
Blackberry, 143
Blog Analysis Toolkit, 93
Bloglines, 93
blogroll links, 72, 93, 130
blogs, 6
    content analysis, 38–9
    ethics, 43, 44–5
    as information networks, 75
    political, 130–2
    as social networks, 63
    web crawlers, 92–3
bonding social capital, 126, 127, 129
boundaries
    network, 51, 70, 72–3, 81–2, 85–6
    system, 55
boundary-spanning, 53
    see also brokerage
bridge-building, 53
    see also brokerage
bridging social capital, 126–7, 129
British Election Study, 30
brokerage
    betweenness centrality, 75–6
    structural holes, **58**, 161

C, **150**
C++, **150**
causality, xvi, 126, 127–8, **128**, 161–2
    reverse, 160
CCDF (complementary cumulative
    distribution function), 134–6, 134t,
    135, *136*, 168
CDF (cumulative distribution function),
    134, 134t, *134*
censorship, 10, 144–6
censorship-resistant currency, **139**
census data, 24
censuses, 26, 36
centralisation, 76–7
centrality
    betweenness, 75–6
    closeness, 76
    and social influence, 106
chance errors, 26
chat rooms *see* threaded conversations
chat software, 32, 34
children, unintended surveying of, 29

China, censorship, 144
citation hyperlink networks, 82, 84–6, 89
citations, post, 130
    see also permalinks
citizen journalists, 14
civic generation, 125
civil society *see* NGOs
civil unrest, 142, 143–4
classical contracts, 151t, 152
clicks, 120
closeness centralisation, 77
closeness centrality, 76
closure, structural holes, **58**, 161
clustering, 108–9, 171
    temporal, 107, 108
co-link analysis, 87
coding schemes, 38, 39
collaboration
    information public goods, 113–14,
        152–6
    peer production, 150, **150**, 151–2
collaborative filters, 173
Collaborative Web *see* Web 2.0
collective behaviour *see* organisational
    collective behaviour
collective identities, 11, 12, 79, 84,
    116–18
commerce *see* marketing; sales
    distribution
commons-based peer production *see* peer
    production
Commonwealth Scientific and
    Industrial Research Organisation
    (CSIRO), 136
communication networks, 74, 75, 106
communities of interest, 11–12, 32
community, concept of, 10–11
community volunteerism, 123, 125
competitiveness, 141
compiled software, **150**
complementary cumulative distribution
    function (CCDF), 134–6, 134t, *135*,
    136, 168
complete networks, **50**, 51, 70, 72, 86
complex contagions, 108, 143
computer science, 14
confirmation, multiple sources of,
    109, 143

confounding factors, 22
connectedness, 142
consent, informed, 29, 33, 34, 43–6
conservatives, 130–2
construct validity, xvi–xvii, 16, 120, 140, 157, 158
consumption capital, **167**
contagion, social *see* social influence
content analysis, 24, 35–40, 43–7
content segmentation, 39
contextual factors, 104, 168–9, 172–3
contractual frameworks, 151–2, 151t
control groups, 22, 108, 170
control mechanisms, 151–2, 151t
convenience sampling, 27, 29, 30
copyright laws, 10, 44n
core networks, **50**, 123, 126–7, 129
correlation, xvi, 161–2
  reasons for, 104
  spurious, 105, 127–8, 159
coverage bias, 26, 30
CSIRO (Commonwealth Scientific and Industrial Research Organisation), 136
cultural markets, 41, 171–2
cultural preference data, 70, 98, 107
cumulative advantage, **122**, 158–9, 172, 173
cumulative distribution function (CDF), 134, 134t, *134*
curator tools, **92**
customer satisfaction surveys, 28
cyberbalkanisation, 129–30, 132
cybermetrics *see* Webometrics
cyberspace, 8–10
cyberspace ethos, 9–10, 144
Cyworld, 7

DARPA (Defense Advanced Research Projects Agency), 1–2
data collection tools, automated, 46
  see also web crawlers
data preparation, 40
data sources and tools
  social network sites, 70, 71
  threaded conversations, 68, **69**
  Twitter, 73
dating, online, 101–3

deep links, 140
Defense Advanced Research Projects Agency (DARPA), 1–2
degree, 75, **122**, 171
  see also indegree
democratising access, **117**
  to political information, 120, 122
  to scholarly expertise, 158–9
depression, 39, 45
diary research, 24
digital currencies, **139**
digital divide, 9
digital trace data, 24, 35–40, 43–7, 112–13
directed edges, 50
discourse analysis, 43
discussion boards
  academic, 158
  self-disclosure on, 39
  SeniorNet, 39, 45
  see also threaded conversations; Usenet
discussion people, 154, *155*, *156*
discussion topics, 66–8, *67*, *68*
diversity, of core networks, 126–7, 129
DNS (Domain Name System), **3**, 5
domain names, **3**
dyads, **49**, 56, 75, 86, 102

e-Government, 138
early adopters, 108
economics of superstars, **166**
edge lists, 52, 52t
edges, 48, 50
  see also ties
efficiency, 142
egalitarianism, 9
ego networks, **50**, *53*, *54*, 76
  answer and discussion people, 154
  citation hyperlink networks, 86
  from Facebook, 71
  Second Life, 161
eHarmony, 101
election polls, 29–30
email, 2
  online interviews, 31, 32–3
  online surveys, 25, 26, 27–8
  unsolicited, 29
  see also Usenet

email lists, harvested, 28
empirical regularities, 14, **122**
employment contracts, 151t, 152
endogenous network effects *see* purely
    structural network effects
engagement, political *see* political
    engagement
environmental activist websites, 36, 84,
    85, 117, 118
epidemiological research, 146
ERGMs (exponential random graph
    models), 60–1
    homophily, 61, 100, 118
    hyperlink networks, 83, 85, 86, 115, **117**
errors, 26–7, 30, 70
ethics
    digital trace data, 43–7
    harvested email lists, 28
    online interviews, 32, 33
    online surveys, 28–9
    web crawlers, 90–1
ethnography, virtual, 42
EverQuest (EQ), 7, 41, **125**, 147
experiments, 22–3, 105
    online, 40–2, 108–10, 141, 170, 172–3
exponential random graph models *see*
    ERGMs
extrinsic motivation, 152–3
extroversion, 141
eyeballs, 120

face-to-face interviews, 31, 32, 33
Facebook, **4**, 7, 13, 24, 41, 50
    access to datasets, 70
    content analysis, 35, 36, 40
    ethics, 43, 44, 46
    friendship formation, 99–101
    networks, 63, *64*, 69–71, 74
    social influence, 169–70
field experiments, 22, 41, 105,
    108–10, 170
field research, 23, 41–2
filtering *see* censorship
fixed-effects models, **128**
FLIXSTER, 170
focus groups, 23, 33–4
followers, Twitter, 72, 74, 109–10
force-directed graphing, 130n, *131*

foreign office websites, 141
forums, 31, 153–4
    see also threaded conversations
fragmentation, 129–30, 132, 159
frames
    collective identities, 116
    sampling, 26, 30
free rider problem, 113
Free Software Foundation, **150**
freedom, 10, 144
frequency–rank plots, 165
friends, Facebook, 70, 74, 99, 101
friendship formation, 97–9, 99–101, **100**
friendship networks, school *see* school
    friendship networks
Friendster, 7, 101

games, online, 123, **125**
Generation X, 125
geodesic distances, 76
German Socio-Economic Panel Study
    (SOEP), **128**
Germany, elections, 29
gift culture, 153
globalisation research, 114, **115**
Google, **4**, 80, 90, 91, 122–3, 157
    see also search engines
Google Docs, **4**
Google Groups, 68, 153
Googlearchy, 122
governance, of Internet, 5
governance structures, 151–2, 151t
government
    authority, 138, 142–6
    censorship, 144–6
    civil unrest, 142, 143–4
    tools of, 138–9, **139**
government websites, 5, 82
    .gov domain, 141–2
    hyperlink networks, 139–42
graph-theoretic attributes, 49, 141–2
grassroots organisations *see* NGOs
Great Firewall of China, 144
group size, 98, **100**, 132

hacking, 29
Harry Potter, 172
harvested email lists, 28

hashtags, 36, 73
Hawthorne effect, 24
health behaviour, 108–10
Heritrix, **92**
heterogeneity, unobserved, 128, **128**, 160
heterophily, **100**
hierarchy, 142, 151t, 151–2
historical web data, 91
history of Internet, 2, **4**
hits, website, 120
HIV/AIDS, 79, 114
homogeneity, 11, 12
homogeneity index, **100**, 132
homophily
  Facebook friendship formation,
    99, 101
  measurement of, 97–9, **100**
  online dating, 101–3
  online social movements, 117
  political, 129–32
  and social influence, 107–8, 109–10
homophily effects, 59
homophily index, **100**
hostnames, **3**, 81
HTML, 2, 84, 86–7, **88**
HTTP, 2, 84
human capital, **57**
human-subjects research, 43
hyperlink counts regression, 82, 83, 157
hyperlink network analysis, 80–2, 83,
  85, 89
hyperlink networks, 78–94
  citation, 82, 84–6, 89
  government, 139–42
  as information public goods, 113–14
  international, **115**
  issue, 83, 84–6, 89
  NGOs, 113–14, **117**
  nodes, ties and boundaries, 80–2, 84–6
  SMOs, 116–17
  social, 83, 84–6, 90
  web crawlers, 86–7, 86–91
hyperlinks, 5
  in blogs, 72, 93, 130
  collective behaviour, 112–13
  measure of scholarly output, 14, 156–7
  outbound, 79, 82, 89, 90, 141, 142
  see also inbound hyperlinks

HyperText Markup Language *see* HTML
HyperText Transfer Protocol *see* HTTP

ICANN (Internet Corporation for
  Assigned Names and Numbers), 5
IEEE (Institute of Electrical and
  Electronic Engineers), 5
IETF (Internet Engineering Task
  Force), 5
IIPC (International Internet Preservation
  Consortium), 91, **92**
IM (instant messaging), 107–8
imperfect substitution, **167**
inbound hyperlinks, 79–80
  citation hyperlink networks, 82, 85
  collection of, 89, 90
  deep links, 140
  political information, 120, 121–2, **121**
  and power laws, 15, 121–2, **122**, 133–6,
    133–6, 134t
  scholarly output, 14, 156–7
  and search engine ranking, 80, 116
inbreeding heterophily, **100**
inbreeding homophily, **100**
incentive mechanisms, 151–2, 151t
incentives, 30
inclusiveness, network, 76, 142
indegree, 75, 121, **122**, 134, 134t,
  164, 165
  see also inbound hyperlinks
indegree centralisation, 77
indegree–rank plots, 133, *133*, 164, *164*
index authority, 84, 116
influence, 53
  of general public, 171–3
  in markets, 168–73
  of networks, 159–60
  social, 61, 103–10, 168, 169–71, 173
information flows, 74–5
information networks, 74–5, 106, 110,
  138, 140
information public goods,
  113–14, 152–6
information science, 14, 15, 82, 87
informed consent, 29, 33, 34, 43–6
informetrics, 14, 156
Infoscape, 93
inlinks *see* inbound hyperlinks

innovation, 160
instant messaging (IM), 107–8
Institute of Electrical and Electronic
	Engineers (IEEE), 5
instrumental behaviour, 115
instrumental variables, 105, **128**
intelligent personal agents, **4**, 5
intense ties, 154
inter-actor analysis, 89
intercept surveys, 28
interdependence, 11, 55, 56, 74, 83
interlocking directorate approach, 116
International Internet Preservation
	Consortium (IIPC), 91, **92**
International Telecommunications Union
	(ITU), 5
internationalism, 9
Internet Archive, 91, **92**, 132
Internet Corporation for Assigned
	Names and Numbers (ICANN), 5
Internet Engineering Task Force
	(IETF), 5
Internet Protocol (IP), 2
Internet surveys, 25–30
Internet, the
	governance, 5
	history of, 2–3, **4**
	technology, 1–2, 5
interpersonal networks, 48, 49
interpreted software, **150**
interviewer effects, 70
interviews, 23, 31–3, 34–5
intranets, 2n
intrinsic motivation, 153
IP addresses, **3**, 5
iPhones, 9
IQ tests, 16
Iranian demonstrations, 142
isolates, 52, 76
isolation, 123, 129
issue hyperlink networks, 83, 84–6, 87
issue networks, 14–15
IssueCrawler, 87, 89, 90
ITU (International Telecommunications
	Union), 5

joint consumption technology, **167**
jointness of supply, 113–14

laboratory experiments, 22, 40–1
labour markets, 153, 160
	see also Mechanical Turk
latent content, 36–7, 38
latent content analysis, 37
least upper boundedness, 142
LexiURL Searcher, 90, 141
liberals, 130–2
LinkedIn, 7, 69
Linux operating system, **4**, 66, 150
list-based sampling, 27
logistic regression, 108
Long Tail, *133*, 134, 163–6, *164*, *165*,
	166–8, 173
longitudinal research, 70–1, 105, **115**, **128**

macroeconomics, 147
manifest content, 36–7, 38
mapping, 8–9, 146–7
marketing, 28, 33, 168–73
markets, 151t, 152
marriage data, 97, 101
Mars, 151
mass media, 127, 145
massive multiplayer online role-playing
	games (MMORPGs), 7, **125**
matched sampling, 108
Matthew effect *see* cumulative advantage
measurement errors, 27, 30, 70
Mechanical Turk (MT), **42**, 172–3
media studies, 14–15, 83
metrics
	network-level, 70, 76–7
	node-level, 70, 75–6
microblogs, 7, 64–5, *65*, 72–3
	see also Twitter
Microsoft Netscan Project, **69**, 153
minimum paths, 75, 76
mixed-mode surveys, 30
MMORPGs (massive multiplayer online
	role-playing games), 7, **125**
mode effects, 29, 30
modelling
	agent-based, 144
	public policy, 146–7
	see also ERGMs
money, quantity theory of, 147
money supply, 138, **139**, 147

moral panic, 37, 123, **125**, 125
motivation, 152–3
MSN, 157
multimodal networks, **50**, 63
multiplex networks, **50**, 64–5
MySpace, 37–8, 43, 69
MySQL, 150

NameGenWeb, 71
nanotechnology example, 37
narrowcasting, 122
NASA Clickworkers, 151, 152
natural experiments, 23, 41, 105
navigation, 140–1
negative affect relations, 84, 85
NetarchiveSuite, **92**
netiquette, 32, 68
Netscan, **69**, 153
network boundaries, 51, 70, 72–3,
    81–2, 85–6
network definitions, 48–53, **49**
network density, 76
network effects, 59–60, 159–60
network inclusiveness, 76, 142
network-level metrics, 70, 76–7
network science, xv–xvi, 14, 15
network size, 76
network structure, 108–9, 159–62, 171
network visualisation tools, 89–90
New Jersey Income Maintenance
    Experiment, 22
new social movement theory, 115–16
newsgroups, 44, 45–6, 153–4
    see also threaded conversations
NGOs (non-government organisations),
    111–12, 113–14, 116, **117**
    see also SMOs
nodality, 82, 138, 140–1
node attributes, 49–50, 61, 141–2
node-level metrics, 70, 75–6
nodes, 48, 49–50, 75, 80–1, 84
NodeXL, 73, 90
non-excludability, 113, 153
non-government organisations see NGOs
non-graph-theoretic attributes, 49–50
non-list-based sampling, 27
non-participant observation, 23
non-probability Internet surveys, 28

non-probability sampling, 27
non-reactive research see unobtrusive
    research
non-response bias, 26, 28, 30, **42**
non-rivalry, 113–14, 153
non-sampling errors, 26–7, 30, 70
normal distribution, **132–3**, 157, *166*
normalisation thesis, 120, **121**
North–South divide, 114
NutchWAX, **92**

obtrusive research, 24, 25t
Occupy Movement, 142
ONI (OpenNet Initiative), 144–5
online communities see virtual
    communities
online dating, 101–3
online games, 123, **125**
online groups, 11–12
online interviews, 31–3, 34–5
online polls, 29–30
online research methods, 21–47
    focus groups, 33–4
    interviews, 31–3, 34–5
    surveys, 25–30
    web content analysis, 35–40, 43–7
online social networks, 12–13
    see also social networks
online surveys, 25–30
ontologies, **88**
Open Data, 5
open licence contracts, 151t, 152
open source software, 150, **150**, 152–3
openness, 9
OpenNet Initiative (ONI), 144–5
opportunity structures, **58**, 98
ordinary least squares, 56
organisational capacity, government, 138
organisational collective behaviour,
    111–18
    information public goods, 113–14, 152–6
    social movements, 115–18
out-groups, 129, 131
outbound hyperlinks, 79, 82, 89, 90,
    141, 142
outdegree, 75
    see also outbound hyperlinks
outdegree centralisation, 77

outlinks *see* outbound hyperlinks
overblocking, 145

PageRank, **4**, 110
Pajek, 90
panel conditioning, 28
panel data, 105, **128**
partial networks, **50**
participant observation, 23, 43
participation, political, 127–8, **128**
participation, thresholds of, 29
participatory democracy, 129, 132
passive-broadcast referrals, 169–70
peer effects, 23, 104n, 105
peer production, 150, **150**, 151–2
Perl, 150, **150**
permalinks, 72, 93, 130
personal attributes, 97–8, 101, 102, 103
personal information disclosure, 33, 37–8,
     39, 43
personal networks, **50**
     see also ego networks
personalised referrals, 169–70
pervasive awareness, 129
PHP, 150, **150**
physics, applied, xv, 14, 15, 120
political attention, 119–20, 121
political attitude formation, 105
political blogs, 130–2
political engagement, 127–8, **128**
political homophily, 129–32
political information, 119–23, **121**
political party websites, 82
polls, online, 29–30
pop-up surveys, 28
population shares, 98, 101, 132
positive affect relations, 84
positive dimension, 159
post citations, 130
     see also permalinks
power law coefficients, 168
power laws, 14, 15, **132–3**
     development of, **122**
     hyperlink counts, 121–2, 132–6,
          133–6, 134t
     online dating, 102
     sales distribution, 164–6, *164*, *165*
practical resources, exchange of, 84, 116

pre-recruited panel surveys, 28
preferential attachment, 14, **122**, 158, 166
price levels, 147
privacy, 29, 31
private/public distinction, 45–6
probability-based Internet surveys, 27–8
probability-based sampling, 27, 30
probability density functions, 164–6,
     165, 166
product adoption, 107–8, 169–71
produsage, 3
propinquity mechanism, 98
prosumption, 3
protocols, communications, 2, **3**, **4**, 5
proximity, 11, 12, 98
public choice theory, 113
public goods, 113
     information, 113–14, 152–6
public opinion
     focus groups, 33
     ratings systems, 171–3
public policy research, 146–7
public/private distinction, 45–6
public sphere theory, 83
purely structural network effects, 59–60,
     61, 83, 117
Python, 73

qualitative content analysis, 24, 38–9
qualitative research methods, 22, 23, 25t
quantitative content analysis, 24, 35–8
quantitative research methods, 22, 23, 25t
quantity theory of money, 147
quasi-experiments, 23
question-and-answer topics, 66–8, *67*

radio telescope data, 151
random assignment, 23
random digit dialling (RDD), 27
random sampling, 27
ratings systems, 171–3
RDF, **4**, **69**, **88**
reactive research *see* obtrusive research
recall error, 70
receiver effects, 59, 84–5
reciprocated edges, 53
reciprocity, 53, 55, 59, 98–9, 114
recommendations, friend, 42, 172–3

recommender systems, 159, 173
reconfiguring access, xvii, 120, 158–9
recording units, 35–6, 38, 39
recruitment, 32
redundant ties, **57**, 108
referrals, 169–71
regression analysis, 82, 86, 108
relational hyperlink analysis, 83
Renren, 7
reply networks, 67
representativeness, 29–30, 32
research disciplines, 13–15
research methods
 qualitative, 22, 23, 25t
 quantitative, 22, 23, 25t
 see also online research methods
research modes, 22–4, 25t
researcher presence, 24
researcher-subject contact, 24
resource mobilisation, 115
resources, exchange of, 84, 116, 117
response rates, 30
retrievability, 80, 120, 158
retweets, 72, 109–10
reverse causality, 160
rich-get-richer *see* cumulative advantage
risk, participant, 34
robots.txt protocol, 91
RSS feeds, 6
RSVP, 101

sales distribution, 159, 163–6, *164*, *165*,
 167, 167–8, 173
sample unit selection, 26
sampling, xvii, 26–7, 29–30, 35, 36, 38
 matched, 108
 see also snowball sampling
sampling bias, 26, 28, 30, **42**, 145
sampling errors, 26, 30
sampling frames, 26, 30
sampling units, 35, 38
scholarly expertise, access to, 158–9
scholarly output, 14, 156–7
school friendship networks, 51–3, *51*,
 52t, *53*, *54*
 controlling for group size, **100**
 SNA metrics, 75–7
 transitive triads, 59–60, *59*

scientometrics, 82, 156–7
search engines, 5, 29, 90,
 120, 122–3
 and anonymity, 46
 APIs, 91
 differences in results, 157
 ranking, 80, 116
Search for Extraterrestrial Intelligence
 (SETI) project, 151
Second Life, **4**, 7, 40, 160–2
secondary data analysis, 24
seed URLs, 87, 90
selection effects, 104–5
self-disclosure, 33, 39
 see also personal information disclosure
self-organisation, 61
semantic databases, **4**
Semantic Web, **4**, 5, **69**, **88**
Semantically Interlinked Online
 Communities (SIOC), **69**
sender effects, 59, 98
SeniorNet, 39, 45
sensitive information *see* personal
 information disclosure
sensitive topics, 45
SETI@home, 151
sexual partnerships, 101–3
shaping force, web as, xiv, 16–17, 143
shortest path length, 141
signalling, 153
silences, online interviews, 33
Simple Mail Transfer Protocol
 (SMTP), 2
simple random sampling, 27
simplex ties, 65
Sina Weibo, 7
SIOC (Semantically Interlinked Online
 Communities), **69**
sitenames, **3**
Skype, 32, 34
Slashdot, 6
smoking
 policy response, 138–9
 teenager behaviour, 61, 104
SMOs (social movement organisations),
 111, 116–17
 see also NGOs
SMTP (Simple Mail Transfer Protocol), 2

SNA *see* social network analysis
snowball sampling, 27, 36
  hyperlink network research, 82, 86,
      115, 117, 141
  web crawlers, 89, 90
social behaviour, traces of, 24
social capital, **50**, 123, **124**, 125–7, 129
social change *see* organisational collective
  behaviour
social dynamics, 171, 172
social groups, 10, **58**, 99, 126, 129
social hyperlink networks, 83, 84–6, 90
social impact, xiii, xiv, 16–17
  see also moral panic; social unrest
social influence, 61, 103–10, 168,
    169–71, 173
social isolation, 123, 129
social media
  social influence, 105–10
  use by protestors, 142, 143–4
social media networks, 40, 61–73
  microblogs, 64–5, *65*, 72–3
  social network sites, 63, *64*, 69–71
  threaded conversations, 61, *62*, 65–8,
    67, *68*
social movement organisations (SMOs),
    111, 116–17
  see also NGOs
social movements, 115–18
social network analysis (SNA), 37, 48,
    55–61, **57**, 83, 104
  metrics, 70, 75–7
social network sites, 7, 12–13, 43, 84–6
  ethics, 45, 46
  social influence, 169–71
  as social networks, 63, *64*, 69–71
  see also Facebook
social networks, 12–13, 74, 75, 106
  definitions, 48–53, **49**
  nodality, 138, 140
  structural holes in, **58**
  types, **50**
  see also school friendship networks;
    social media networks
social problems, 116–17
social relations, 11, 55–6
social selection, 61, 104, 168
  see also homophily

social structures, 55, **57**
  see also structural holes
social systems, 55
social tool, web as, xiv, 16–17, 143
social unrest, 142, 143–4
sociality, 98, 101
socio-demographic data, collecting, 33
socio-economic data, 70
sociograms, 48
SocSciBot, 89–90
solicited diaries, 24
source code, **150**
SPARQL, **4**
Spinn3r, 93
Static Web *see* Web 1.0
static websites *see* Web 1.0 websites
statistical inference, 26, 27
statistical-mechanical models, 14, 15
strategic advantages, **58**, 159
stratified random sampling, 27
strength of weak ties, **57**, 108–9, 126
strong ties, 126
strongly connected networks,
    76, 140–1
structural holes, **58**, 118, 126, 161–2
structural signatures, 16, 117, 153–4
stylised facts, 14, 172
subdomains, **3**, 81
subsites, **3**, 81
superstars, 164, 166, **167**, 172
surveillance *see* censorship
surveys, 23, 25–30
SWRL, **4**
symbolic resources, exchange of,
    84, 116
synchronous focus groups, 34
synchronous online interviews, 31, 32
systems, 55

target populations, xvii, 26
taxation, 138
TCP/IP, 2, **4**
telephone surveys, 25, 29
temporal clustering, 107, 108
text content, 35
text mining, 122, 130
text parsing, 68
thread networks, 61, *62*, 65–8, *68*, 75

threaded conversations, 6, 153–4
  as social networks, 61, *62*,
    65–8, *67*, *68*
thresholds of participation, 29
tie formation behaviour, 55, 61, 74
ties, 48
  hyperlink networks, 81, 84–5
  instense, 154
  redundant, **57**, 108
  simplex, 65
  strength of weak ties, **57**, 108–9, 126
  strong, 126
  weak, 129, 143, 160
  see also edges
TLDs (top-level domains), **3**, 5
tools of government, 138–9, **139**
top-level reply networks, 67
transaction costs, 151
transcripts, automatic, 32
transitive triads, 53, 59–60, *59*
transitivity, 53, 55, 59, 98–9
Transmission Control Protocol (TCP), 2
treasure, tool of government, 138, **139**
treatment groups, 22, 108, 170
triadic closure, 53, 60, 101
trust, 123, 125, 161
tweepy, 73
tweets, 36, 72–3
Twitter, **4**, 7, 24, 49
  API, 73
  content analysis, 35, 36, 40
  as information network, 74, 75
  networks, 50, 64–5, *65*, 72–3
  social influence, 107, 110
  use by protestors, 142

undirected edges, 50
uniform resource locators *see* URLs
unimodal networks, **50**, 63
United Kingdom
  censorship, 144
  elections, 30
  government websites, 141
  riots, 143
United States, government websites, 141
units of analysis, 35–6, 39, 86
universalism, 9
unobserved attributes, **124**, 160

unobserved heterogeneity, 128, **128**, 160
unobtrusive research, 24, 25t
  see also content analysis
unrestricted self-selected surveys, 28
unsolicited diaries, 24
  see also blogs
unweighted edges, 51
URLs, **3**
  seed, 89, 90
Usenet, 6, 68, **69**, 153–4
user navigation experience, 141

viral marketing, 169–70
viral product design, 169–70
virtual communities, 10–12, 13, 41, 129
virtual communities of interest,
    11–12, 32
virtual ethnography, 42
Virtual Observatory for the Study of
    Online Networks (VOSON),
    82, 90, 117
virtual worlds, 7, 41, 160–2
  EverQuest, 7, 41, **125**, 147
  modelling public policy, 146–7
  as social networks, 65
  World of Warcraft, 7, 41, 65, 126, 146
visibility, 80, 120, 140, 158
  political information, 119–20, **121**
  scholarly output, 156–7
visits, website, 120
visualisation, 8–9, 89–90
VOIP (Voice over Internet Protocol), 32
voluntary organisations *see* NGOs; SMOs
volunteerism, 123, 125
VOSON (Virtual Observatory for the
    Study of Online Networks),
    82, 90, 117
vulnerability, participant, 34

W3C (World Wide Web Consortium), 5
walled gardens, 9
Wayback Machine, 91, **92**
weak ties, 129, 143, 160
  strength of, **57**, 108–9, 126
Web 1.0, 2, **4**, 5
Web 1.0 websites, 6, 36, 62, *63*
Web 2.0, 2, **4**, 5, 8, 9
Web 3.0 *see* Semantic Web

web browsers, **4**
web content analysis, 24, 35–40, 43–7
web crawlers, 46, 82, 86–91, 141
  blogs, 92–3
  ethics, 90–1
Web Curator Tool (WCT), **92**
web surfing, 5, 126
web, the
  governance, 5
  history of, 2–3, **4**
  technology, 1–2, **3**, 3, 5
Webometric Analyst, 90
webometrics, 14, 82, 84, 120, 156–7
website hits, 120
website log files, 120
website usage patterns, 14

weighted edges, 51
WERA (WEb aRchive Access), **92**
Wikileaks, **139**
Wikipedia, 6, 73, 150
wikis, 6, 63, *64*
winner-takes-all *see* cumulative advantage
withdrawal, 33
word-of-mouth marketing, 169–71
Wordpress, 43
World of Warcraft (WoW), 7, 41, 65,
  126, 146
world-systems theory, **115**
World Wide Web Consortium
  (W3C), 5

Yahoo, 107, 157